Practical Linux Programming:

Device Drivers, Embedded Systems, and the Internet

Practical Linux Programming:

Device Drivers, Embedded Systems, and the Internet

Ashfaq A. Khan

CHARLES RIVER MEDIA, INC.
Hingham, Massachusetts

Acquisitions Editor: Brian J. Sawyer
Production: Publishers' Design & Production Services
Cover Design: The Printed Image

CHARLES RIVER MEDIA, INC.
20 Downer Avenue, Suite 3
Hingham, Massachusetts 02043
781-740-0400
781-740-8816 (FAX)
info@charlesriver.com
www.charlesriver.com

This book is printed on acid-free paper.

Ashfaq A. Khan. *Practical Linux Programming: Device Drivers, Embedded Systems, and the Internet.*
ISBN: 1-58450-096-4

Library of Congress Cataloging-in-Publication Data

Khan, Ashfaq A.
Practical Linux programming : device drivers, embedded systems, and the Internet / Ashfaq A. Khan.
p. cm.
ISBN 1-58450-096-4 (pbk. : acid-free paper)
1. Linux. 2. Operating systems (Computers) I. Title.
QA76.76.O63 K497 2002
005.4'32—dc21
2001006986

Printed in the United States of America
02 7 6 5 4 3 2 First Edition

CHARLES RIVER MEDIA titles are available for site license or bulk purchase by institutions, user groups, corporations, etc. For additional information, please contact the Special Sales Department at 781-740-0400.

Requests for replacement of a defective CD must be accompanied by the original disc, your mailing address, telephone number, date of purchase and purchase price. Please state the nature of the problem, and send the information to CHARLES RIVER MEDIA, INC., 20 Downer Avenue, Suite 3, Hingham, Massachusetts 02043. CRMís sole obligation to the purchaser is to replace the disc, based on defective materials or faulty workmanship, but not on the operation or functionality of the product.

Dedication

I dedicate this book to my mother, Ashraf Mumtaz.

Table of Contents

Acknowledgments

First, I want to thank Brian Sawyer for giving me the opportunity to write this book—his kind encouragement and trust kept me going. Special thanks to Parham Kamali and Stephanie Kwok for being great friends and providing me help when I needed it. I also want to thank Stephano Stepan, Rojelio Rodriguez, Saleem Qureshi, and Kaleem Abdullah for helping me prepare the design projects presented in the book.

Preface

This book is about designing embedded systems, using Internet technology as a user interface and a base operating system that is real time in nature for a system interface.

Perhaps the distribution of workload as user interface design and system interface design is arbitrary, but it seems natural in knowing that we, as users, are relying more and more upon the Internet for our insatiable appetite for information. This book attempts to incorporate the technology behind the Internet into everyday products that so far have not taken the path of mainstream media. This goal might not have been easy to achieve if it were not for the GNU/Linux platform, which provides an operating system environment that is ideally suited for our purposes. There are multitudes of operating systems available, suitable for real-time applications—we have chosen Linux. The criteria for selecting Linux include cost, portability, operability, sustenance, ease of use, and reliability, to name a few. Actually, there is no better solution than the GNU/Linux platform for taking on a project design of any magnitude.

Just what is an embedded system? We could probably say that an embedded system is a computer dedicated to a specific application, or that a predefined application is embedded in a computer system for a dedicated task. Whatever definition you choose, the concept of an embedded system begins with a specific requirement, probably a marketing decision that fulfills a specific need of a customer or customers. As a system designer, an engineer is faced with the task of fulfilling the marketing requirement of the company; however, the most important factors are usually the cost of the end product and the time to complete the project. There is always a demand to reduce the cost of the product, and a push to finish the project in record time. Engineers always find that "experience is something we have *after* we need it." The pace of change in technology is such that any design you start today may be obsolete before it reaches the market. So, how can you efficiently design and produce a product that is both cost effective and meets the requirement? The answer lies in using the technology that has already been developed, instead of reinventing the wheel. The modules that form the backbone of the Internet are available to us, and there is no better way to learn than from the experience of others. We only have to know how to use the existing modules instead of designing a new one. In fact, nowadays, a software engineer spends more time learning the interaction among the existing modules than designing a new one from scratch.

The challenge of a computer system design from an embedded system designer's point of view is both acquiring the knowledge of the hardware and integrating the different modules of the software. It is especially true for a small-scale embedded system where a simple micro-controller is used and the total engineering force is comprised of one individual. However, things become complicated when your system requires interaction between a database, an Internet connection, a network communication, a graphical user interface, and probably a stringent requirement of system integrity. The general guidelines for designing a system such as this are selecting an operating system that has a proven record of robust performance, availability of plenty of support software, and the freedom to choose any component of the system without paying exorbitant royalties to the provider. The only operating system that meets all these criteria is Linux, and the purpose of this book is to give you enough insight into the system to make you feel comfortable with designing a system around it.

The book emphasizes the use of three different technologies for embedded system design and development: the Web, the Linux kernel, and SQL queries. The Web provides the user interface, the Linux kernel provides the system interface, and the SQL database server provides the glue logic between the user and the system interface. From a software design point of view, we cover in detail device driver design, interprocess communication usage, Perl programming, shell programming, HTML tags, and SQL queries. The examples presented in this book cover only a few applications, but the essential theme is to demonstrate the guidelines for designing an embedded system that requires interaction of different software modules, and show you how an operating system like Linux helps you glue your software modules together.

HOW THIS BOOK IS ORGANIZED

The book is presented as a tutorial for students and engineers who wish to learn the process of designing an embedded system application using Linux as the real-time operating system and the Internet as the user interface.

ON THE CD

Chapter 1, "Linux Onboard," discusses Linux installation from the ground up, and includes the steps involved in a full server implementation. A minimum Linux setup that fits in a floppy-only configuration is also provided on the companion CD.

Chapter 2, "The User Interface," provides the basics of Internet technology that we will be using for our user interface. The setup includes a DNS server, SQL server, and Web server. You will be able to launch "Welcome" Web pages from your own Web hosting site. This setup is used in the design projects as part of the user interface design for our embedded systems.

Chapter 3, "Device Drivers," is the device driver design, showing you how to establish communication with the external device that is the object of our embedded system design.

Chapter 4, "Tasks and Interprocess Communication," covers the interprocess communication facilities in Linux, including forks, pipes, FIFO, semaphores, message queues, shared memory, and sockets. These are the facilities of the operating system that an embedded system needs to establish a client/server mechanism that is essential for a modular design.

Dynamic Web pages are essential for real-time display of information on the Internet, and they are only possible if we have an interpretive language such as Perl to generate the common gateway interface (CGI) scripts. Perl programming is discussed in Chapter 5, and "Structured Query Language" in Chapter 6.

The scripts to start and stop various Linux services are all written as shell scripts. Chapter 7, "Shell Script Programming," discusses shell programming.

Although hardware math coprocessors are commonplace now, some embedded designs cannot afford the extra cost. A unique implementation of transcendental functions discussed in Chapter 8, "Fixed-Point Arithmetic and Transcendental Functions," is sure to improve the performance of your software math emulation.

Chapter 9, "Embedded System Design Projects," presents hands-on experience in implementing real-life embedded design projects to give you the confidence that the method and the design philosophy being pursued is truly an achievable goal. The goal is to use the Internet as a media for the user interface, the Linux platform as the basis for system interface, and a database as the glue logic to hold the two interfaces together.

Appendices for HTML tags, C language development tools, and miscellaneous Linux services are provided for easy reference.

Knowledge of C programming is necessary to do the examples provided in the book. Experience with the Linux operating system is a big plus, but the book provides all the guidelines necessary to set up the projects. The parallel port of the PC is used as the external stimulus, so no extra hardware is necessary to work through the examples presented in the book.

1 Linux Onboard

In this chapter

- System Requirements
- Linux Installation
- Linux Boot Process
- Linux in an Embedded System

INTRODUCTION

This chapter discusses installing Linux as both a software development tool and a platform for the design and implementation of an embedded product. An embedded system software developer essentially needs two sets of machines; one is the target, and the other is for development. One great advantage of using Linux as an operating system is the straightforward process of porting code from the development machine to the target machine. It is possible to use the target machine for development purposes as well, but it might not be feasible in most cases. Space and feature constraints could slow the development process. It is always preferable to work with a fast machine that has quick compile, test, and debugging turnaround times.

The requirements for installing Linux on a target machine are usually different from those of the development machine. In this chapter, we discuss the Linux installation process from the development point of view, and a minimal Linux system that is suitable for a low-cost embedded system. The choice of Linux distribution is left up to the user, but the Red Hat distribution is by far the most popular. While we make no attempt to recommend any specific hardware, a good development machine should have a 20GB hard disk, 256MB RAM, P3 or above CPU, and the usual peripherals of CD Read/Write, 100MB Ethernet card, and a video card.

The choice of hardware for your embedded Linux is a different matter. It would be nice if the machine you are going to sell looks like the machine on which you developed your software. However, marketing probably has other plans; you'll be lucky if you are allowed a 4MB flash card to store your Linux, and a 32MB RAM SIM to start your Linux. Having said that, the main purpose of this chapter is to guide you through the intricate details of the Linux boot process, and to help you install and port Linux on your embedded machine. To begin, you first need to install Linux from a distribution CD-ROM, such as Red Hat, Mandrake, Debian, SuSE, and so forth. A list of Linux distributions is available at www.linuxdoc.org/HOWTO.

This chapter is divided into two sections. The first discusses the installation procedure for the development machine, where we will install a complete Linux system from a distribution CD. We will use the development machine for compiling, linking, and executing the software.

The second section describes the Linux boot process, and how to port Linux on an embedded machine. You can reduce your time to market if you can simulate most of your end-product system requirements (such as the boot disk) on your development machine. A good simulation environment is essential for a rapid development cycle. There is a learning curve associated with developing software in a Linux environment. However, once you understand the tools and are comfortable using them, you can take on any software development challenge, regardless of how complex it might be. You can also try the newer Integrated Development Environment (IDE) packages such as Kdevelop 1.4, freely available from open source contributors—download it from www.kdevelop.org.

We begin this discussion with the installation of Linux as a development machine.

SYSTEM REQUIREMENTS

The Linux installation and setup has come a long way from its cumbersome early days. Several distribution CDs now offer quick and fairly easy installation scripts. The latest trend in Linux distribution is toward auto-installation and auto-detection of the system hardware. Red Hat Linux can be installed on almost all hardware developed in the last two years. The examples in this book were written using Red Hat Linux 7.1 and are compatible with version 7.2. For information on downloading and installing Red Hat Linux, go to www.redhat.com/download/howto_download.html.

You will need an IBM PC compatible computer with a Pentium class CPU, a minimum of 32MB RAM, and at least a 2GB hard drive for the Linux installation. The requirement for Red Hat version 7.1 server-class is a minimum of 1.2GB of free disk space, plus swap space (twice the amount of system RAM). Proper partitioning of the hard disk is important, because we want to create an extra partition to simulate the embedded system design disk requirement on our development machine. Several utilities are available for partitioning a hard disk, but fdisk is the most flexible and offers a variety of file system types. We discuss the utility in detail as we proceed with the Linux setup.

Supporting Multiple Operating Systems

Linux and Windows can coexist peacefully on the same hard drive, which might be a good solution for most users. If your existing hard drive has only one partition and you want to preserve the installed Windows 9x, you need to run the *defrag* program of Window 9x and use the DOS-based *fips* program. Follow the online instructions when you run *fips* to create or add a new partition. Linux requires a minimum of two partitions on the hard disk, one primary and one extended. The extended partition must be further divided into partitions before logical drivers can be created. If you plan to include Windows 98 as well, use the primary partition for Windows 98, and one of the extended partitions for Linux. Linux requires a minimum of two partitions.

Swap Partition

The swap partition is used mainly for swapping the memory contents during large RAM access. The maximum size for this partition is two times the amount of RAM being used.

Root Partition

Once the swap space is determined, the rest of the disk can be allocated for the root file system. However, we recommend having multiple partitions on the hard disk so that you can try different file systems on the same drive. If your system has 32MB RAM and a 2MB hard disk, you will need 64MB of swap, and the rest of the disk space for the Linux native file system. Linux uses a hierarchical file system where every resource is a file, and all files on the system fall under the root file system "/". Devices such as CD-ROMs and disk partitions are named like any other file on the system, but need to be mounted before they can be used. The process of mounting a device essentially tells Linux what type of file system should be used to access the contents of the device. Hard disk devices are named *hda* for disk 1, *hdb* for disk 2, and so on. On each disk, you can have one primary partition hda1 and up to three extended partitions—hda2, hda3, and hda4. (Only use hda1 and hda2, because

DOS recognizes only one primary and one extended partition.) The extended partition needs to be divided into logical drives—starting at hda5, hda6, and so on—before they can be used as devices to mount file systems. The Red Hat CD offers two utilities—Disk Druid and fdisk—to perform disk partitioning. Disk Druid is more user friendly, but fdisk offers more choices of installing different file systems. We discuss fdisk later as we proceed with the installation.

LINUX INSTALLATION

The most efficient media for Linux installation is the CD, but you can also use a floppy disk, which we discuss later in the chapter.

Installing from the CD-ROM

Most motherboards have a CMOS setup option that allows you to boot from a CD-ROM directly. This is usually activated by depressing the DEL key while the system is rebooting. Once in CMOS setup, you will see several suboption menus. Select the one that allows you to reconfigure the default boot device to be the CD-ROM. This option is usually found in the Advanced Features submenu. After selecting the option, update the CMOS setting, exit, and let the boot process continue (make sure your Red Hat CD is in the CD-ROM drive).

Restart the machine and let the boot process begin.

Installation Options and System Requirements Table

You will be presented with a series of windows and will be asked to select different options. Refer to Table 1.1 for the server-class setup of Red Hat version 7.1. We recommend that you perform a manual partitioning of the hard disk using the fdisk utility, as it is very flexible and supports a variety of file systems.

fdisk

The fdisk utility is the preferred tool for partitioning the hard disk. You can use fdisk after the system is installed, but be very careful not to disturb the partitions that contain vital data or other operating systems you want to preserve. As mentioned previously, you need a minimum of two partitions, Linux native and Linux swap. However, if your embedded system design has a Linux boot option with a flash file system, it would be worthwhile to add another partition of the same size as that from which your embedded design is supposed to boot. Doing so will allow you to test your finished product directly from your development machine. The boot process will execute the fdisk utility and open a shell for you if you choose manual partitioning.

TABLE 1.1 The Screen Reference for the Server-Class Setup of Red Hat Version 7.1

Windows you see and Commands and Responses		
WINDOW	**PROMPT**	**RESPONSE**
Welcome	Boot	<return>
Red Hat Linux	OK	<return>
Choose a Language	English	<return>
Keboard Type	US	<return>
Installation Method	Local CD-ROM	<return>
Note	OK	<return>
Installation Path	Install	<return>
Installation Class	Custom	<return>
SCSI Configuration	No	<return>
Disk Setup	FDISK	<return>
Partition Disk	Edit	<return>

Press *m* <Return> to see the Help menu, and familiarize yourself with some of the commands, especially *d,n,p,t,* and *w*. Press *p* <Return> to view the current partition.

- **d** Delete an existing partition
- **p** Print the partition table
- **n** Add a new partition, or create a primary and secondary partition
- **n-1** Add logical drives to the secondary partition
- **t** Change a partition's file system type
- **w** Update the changes

NOTE

There can only be one primary and one secondary partition. Only the secondary partition can be subdivided into logical drives.

fdisk will help you partition the hard drive, but we need to discuss several options first.

- If you want to make a dual-boot machine with Windows 9x and you already have Windows installed, you should never overwrite the primary partition.
- If you have a new hard drive and you want to make a dual-boot machine, you must create one primary and at least two extended partitions.
- If you only need Linux, you can simply create one primary and one extended partition.

Using fdisk to Delete Partitions from a Used Hard Drive

1. Press *m* <Return> to see the Help menu, and familiarize yourself with some of the commands, especially *d*, *n*, *p*, *t*, and *w*.
2. Press *p* <Return> to view the existing partition table. It should look similar to Figure 1.1.

```
Command (m for help): p

Disk /dev/hda: 255 heads, 63 sectors, 525 cylinders
Units = cylinders of 16065 * 512 bytes

   Device Boot    Start       End    Blocks   Id  System
/dev/hda1   *         1       192   1542208+   6  FAT16
/dev/hda2           193       525   2674822+   5  Extended
/dev/hda5           193       384   1542208+  83  Linux
/dev/hda6           385       508    995998+   6  FAT16
/dev/hda7           509       525    136521   82  Linux swap

Command (m for help): d
Partition number (1-7): 7

Command (m for help): d
Partition number (1-6): 6

Command (m for help): d
Partition number (1-5): 5

Command (m for help):
```

FIGURE 1.1 Screen shot of the fdisk utility *p* command showing the partition table of the hard disk.

3. In Figure 1.1, 7 is the last partition number, since hda7 is the last logical drive; /dev/hda6 corresponds to 6; /dev/hda5 corresponds to 5; and so on.
4. Press *d* <Return> to delete the partition.
5. If the last logical drive is /dev/hda7, enter *7* <Return>, or whatever the last logical drive number is (e.g., for /dev/hda6, enter *6*; for /dev/hda5 enter *5*, and so on).
6. Repeat step 5 until all logical drive partitions are deleted.
7. Delete the extended partition, /dev/hda2, by pressing *d* <Return>, and enter *2*.

Caution:

Do not add or delete the primary partition if you are planning to preserve your Windows 9x.

8. Proceed to step 9 if you want to preserve the existing Windows partition. Delete the primary partition, /dev/hda1, by pressing *d* <Return>, and enter *1*.
9. Press *p* <Return>to view the partition table. Figure 1.2 shows the deleted partition numbers and the blank disk after deleting all partitions.

```
Disk /dev/hda: 255 heads, 63 sectors, 525 cylinders
Units = cylinders of 16065 * 512 bytes

   Device Boot    Start       End    Blocks   Id  System
/dev/hda1   *         1       192   1542208+   6  FAT16
/dev/hda2           193       525   2674822+   5  Extended

Command (m for help): d
Partition number (1-5): 2

Command (m for help): d
Partition number (1-4): 1

Command (m for help): p

Disk /dev/hda: 255 heads, 63 sectors, 525 cylinders
Units = cylinders of 16065 * 512 bytes

   Device Boot    Start       End    Blocks   Id  System

Command (m for help): 
```

FIGURE 1.2 The *p* command shows a blank disk after deleting all partitions.

fdisk to Preserve the Windows 9x Partition

1. Refer to Figure 1.1 to see the existing partition on a hard drive.
2. Press *d* <Return> to delete the existing partition, starting from the last partition number. Do *not* delete partition 1. /dev/hda1 is partition 1, /dev/hda2 is partition 2, and so on. The logical drives start at partition 5.

Caution:

Do not add or delete the primary partition 1 (/dev/hda1/) for Windows 9x.

3. Press *p* <Return> to view the partition table. You should see the existing primary partition.

Using fdisk to Add the Primary Partition

1. This step is required only if you have a blank disk. Press *n* <Return> to add a new partition. (If you want to preserve the Windows 9x partition on a used hard drive, proceed to the next section.)
2. Press *p* <Return> to add the primary partition.
3. Enter *1* <Return> to accept the first cylinder 1.
4. Calculate the number of blocks you want to reserve for the primary partition (typically, each partition is 8MB). When prompted to enter the last cylinder, enter the number you just calculated; for example, enter *200* for a 1600MB primary partition (Figure 1.3).

Using fdisk to Add an Extended Partition

1. Press *n* <Return> to add a new partition.
2. Press *e* <Return> to add an extended partition.
3. Enter *2* for the extended partition.
4. Enter the number for the default first cylinder, and press *Return*.
5. Enter the number for the default last cylinder, and press *Return*. (This step allocates the remaining hard drive for the extended partition, which you will subdivide into logical drives.) See Figure 1.3.

Adding the Linux Swap Partition as Logical Drives

1. Calculate the size of the swap space you need; it should be two times the amount of RAM. For example, a 64MB RAM system needs 128MB swap space, or 16 blocks (8MB/block).

```
Command (m for help): p

Disk /dev/hda: 255 heads, 63 sectors, 525 cylinders
Units = cylinders of 16065 * 512 bytes

   Device Boot    Start       End    Blocks   Id  System

Command (m for help): n
Command action
   e   extended
   p   primary partition (1-4)
p
Partition number (1-4): 1
First cylinder (1-525, default 1):
Using default value 1
Last cylinder or +size or +sizeM or +sizeK (1-525, default 525): 200

Command (m for help): n
Command action
   e   extended
   p   primary partition (1-4)
e
Partition number (1-4): 2
First cylinder (201-525, default 201):
Using default value 201
Last cylinder or +size or +sizeM or +sizeK (201-525, default 525): 525

Command (m for help):
```

FIGURE 1.3 The command and response for adding a primary and extended partition.

2. Press *n* <Return> to add a new partition.
3. Press *l* <Return> to add logical drives.
4. Enter the number for the default first cylinder, and press *Return.*
5. The last cylinder number should be the first cylinder plus the swap space. For example, your first cylinder number is 200, and your swap space is 16 blocks. The last cylinder number is 216. Enter this value when prompted for the last cylinder number.

Adding the Linux Native Partition as Logical Drives

6. Press *l* <Return> to add logical drives.
7. Enter the first cylinder number.
8. Enter the default last cylinder number (you can enter a smaller number and repeat step 6 to create several logical drives).
9. Now you have a hard disk with a primary and an extended partition. The extended partition is further divided into two partitions with logical drives as shown in Figure 1.4.

```
Command (m for help): n
Command action
   l   logical (5 or over)
   p   primary partition (1-4)
l
First cylinder (201-525, default 201): 201
Last cylinder or +size or +sizeM or +sizeK (201-525, default 525): 497

Command (m for help): n
Command action
   l   logical (5 or over)
   p   primary partition (1-4)
l
First cylinder (498-525, default 498):
Using default value 498
Last cylinder or +size or +sizeM or +sizeK (498-525, default 525):
Using default value 525

Command (m for help): p

Disk /dev/hda: 255 heads, 63 sectors, 525 cylinders
Units = cylinders of 16065 * 512 bytes

   Device Boot    Start       End    Blocks   Id  System
/dev/hda1             1       200   1606468+  83  Linux
/dev/hda2           201       525   2610562+   5  Extended
/dev/hda5           201       497   2385621   83  Linux
/dev/hda6           498       525    224878+  83  Linux

Command (m for help):
```

FIGURE 1.4 The command and response for adding extended partition logical drives.

Using fdisk to Change the Partition Type for Win95

1. Proceed to step 4 if you want to preserve the existing Win95 partition type.
2. Press *t* <Return> to change the partition type; enter *1* for hda1.
3. Enter *b* <Return> for Win95 type.

Using fdisk to Change the Partition Type for Linux Swap

4. Press *t* <Return> to change the partition type for Linux swap. Enter *5* for hda5, or whatever logical drive you reserved for swap space.
5. Enter *82* for Linux native.
6. Press *p* to verify the partition table type of hda1=b, hda2=5, hda5=83, and hda6=83 as shown in Figure 1.5.
7. Press *w* to write back the changes. This will exit fdisk.
8. You should now be back to the Partition Disk window. Select *Done*, and press *Enter*.

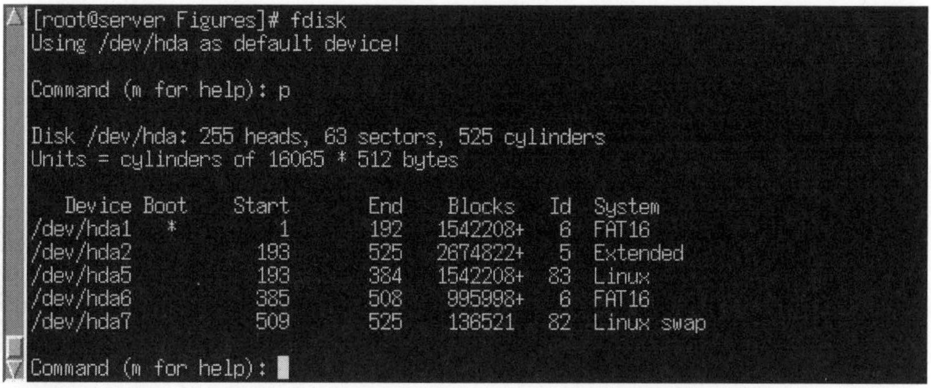

```
[root@server Figures]# fdisk
Using /dev/hda as default device!

Command (m for help): p

Disk /dev/hda: 255 heads, 63 sectors, 525 cylinders
Units = cylinders of 16065 * 512 bytes

   Device Boot    Start       End    Blocks   Id  System
/dev/hda1   *         1       192   1542208+   6  FAT16
/dev/hda2           193       525   2674822+   5  Extended
/dev/hda5           193       384   1542208+  83  Linux
/dev/hda6           385       508    995998+   6  FAT16
/dev/hda7           509       525    136521   82  Linux swap

Command (m for help): 
```

FIGURE 1.5 The fdisk utility *p* command showing the final partition table.

Components to Install

Select to install those components preceded by an asterisk "[*]".

[*] Printer Support
[*] X Windows System
[*] GNOME
[*] KDE
[*] Mail/WWW/News Tools
[*] DOS/Windows Connectivity
[*] File Managers
[*] Graphics Manipulation
[ ] Console Games
[ ] X Games
[*] Console Multimedia Support
[*] X Multimedia Support
[*] Networked Workstation
[*] Dialup Workstation
[ ] News Server
[ ] NFS Server
[*] SMB (Samba) Connectivity
[ ] IPX/NetWare Connectivity

[*] Anonymous FTP Server
[*] Web Server
[*] DNS Name Server
[*] Postgresql (SQL) Server
[*] Network Management Workstation
[ ] TeX Document Formatting
[*] Emacs
[*] Emacs with X Windows
[*] C Development
[*] Development Libraries
[*] C++ Development
[*] X Development
[*] GNOME Development
[*] Kernel Development
[*] Extra Document
[ ] Everything

Installing Devices

Devices on the system include the keyboard, mouse, video card, monitor, network card, modem, and so forth. Red Hat 7.1 can detect devices without user intervention, but version 6.0 might require some manual configuration.

Linux will have no problem detecting a serial or PS/2 mouse, but you will be asked to identify the type of mouse in a new Configure Mouse window. If you do not know the exact type, simply select *Generic type,* and click *OK.*

The Network Configuration window appears next. Click *No,* because you will be using the Control Panel for this selection. Following that is the Configure Time Zone window. Select your local time zone, and click *OK.*

Installing Services

Of the following, select to install those services preceded by an asterisk [*] to be started at machine boot time. Make sure that DNS, Web, and HTTP server are included in the list.

[*] apmd
[*] atd

[*] crond
[*] gpm
[*] httpd
[*] inet
[*] keytable
[*] linuxconff
[*] lpd
[*] mars-nwe
[*] named
[*] netfs
[*] network
[ ] nfs
[ ] pcmcia
[*] portmap
[*] postgresql
[*] random
[*] routed
[ ] rstatd
[ ] rusersd
[ ] rwhod
[*] sendmail
[*] smb
[ ] snmpd
[*] syslog
[ ] ypbind

Other Installation Options

Installing from a Bootable Floppy Disk

If your motherboard does not allow you to boot from a CD-ROM, you must prepare a bootable floppy disk, either from a DOS machine or from an existing Linux system.

Preparing a Boot Floppy Using a DOS/Windows System

1. Insert the Red Hat installation CD into the CD-ROM drive, and a blank floppy into the floppy drive. Execute the program *F:\\Dosutil\rawrite boot.img* (assuming your CD-ROM drive is F).

This will create a bootable floppy disk with boot.img on it. You might need additional files if you are planning to use notebook PCs with built-in PCMCIA support. Execute the *rawrite* program with the option *F:\\Dosutil\rawrite pcmcia.img* (assuming your CD-ROM drive is F).

Preparing a Boot Floppy Using an Existing Linux System

1. Insert the Red Hat installation CD into the CD-ROM drive, and a blank floppy into the floppy drive.
2. Mount the CD-ROM onto the usual CD-ROM mount point, and change the directory using the following shell command:

```
#mount /dev/cdrom -t iso9660 /mnt/cdrom
#cd /mnt/cdrom/images
```

3. Create a bootable floppy disk with boot.img on it with the disk duplicate command:

```
#dd if=boot.img of=/dev/fd0 bs=1440k
```

You might need additional files if you are planning to use notebook PCs with built-in PCMCIA support.

4. Use the following disk-duplicate command to create the supplementary disk:

```
#dd if=pcmcia.img of=/dev/fd0 bs=1440k
```

5. Insert the boot floppy, reboot the system, and follow the procedure as outlined previously in the section *Linux Installation.*

LINUX BOOT PROCESS

All computer systems start the boot process by executing the code in ROM (specifically, the BIOS). The BIOS initializes the basic hardware of the system and prepares the system to boot from one of the boot devices, such as the floppy, the hard disk, CD-ROM, flash card, and so forth. Your embedded system boot process should not be any different. The BIOS reads the code that resides in sector 0 of cylinder 0 of the boot drive you selected, and tries to execute the code.

On most bootable disks, sector 0 of cylinder 0 contains either the code from a boot loader such as "Lilo"—which locates the Linux kernel, loads it, and executes it—or the start of an operating system kernel, such as Linux or DOS.

You can place a compressed Linux on the disk, but the boot sector should have the code that loads the rest of the Linux and decompresses it into RAM. When completely loaded, the kernel initializes device drivers and its internal data structures. Once completely initialized, the kernel consults a special location in its image called *ramdisk*. ramdisk tells it how and where to find its *root file system*. A root file system is simply a file system that will be mounted as "/". Remember, Linux does not use drive letters such as C: or D:. The "/" is the beginning of all file systems; in other words, the root. All other file systems must be mounted under the umbrella of "/" in a separate directory. You must tell the kernel where to look for the root file system; if it cannot find a loadable image there, it halts.

Once the root file system is loaded and mounted, a message similar to the following will appear:

```
VFS: Mounted root (ext2 filesystem) readonly.
```

init

Once the system has loaded the root file system successfully, it tries to execute the init program (in /bin or /sbin).

inittab

init reads its configuration file /etc/inittab and looks for a line containing *sysinit*. Different vendors place sysinit in different directories. The following line is from the inittab of Red Hat version 7.1:

```
si::sysinit:/etc/rc.d/rc.sysinit
```

sysinit

sysinit shows the sysinit script to be found in /etc/rc.d/rc.sysinit. This script is a set of shell commands that set up basic system services, such as running fsck on hard disks, loading necessary kernel modules, initializing swapping, initializing the network, and mounting disks mentioned in /etc/fstab.

sysinit invokes various other scripts to do modular initialization. The directory path for scripts is calculated by the run level; for example, the following line indicates "execute the scripts found in /etc/rc.d/rc5.d":

```
id:5:wait:/etc/rc.d/rc 5
```

When the sysinit script finishes, the control returns to init, which then enters the default run level, specified in inittab with the *initdefault* keyword.

```
id:5:initdefault:
```

The run-level line usually specifies a program such as getty, which is responsible for handling communication through the console and ttys. The getty program prints the familiar "login:" prompt, and in turn invokes the login program to handle login validation and set up user sessions.

```
5:2345:respawn:/sbin/mingetty tty5
```

LINUX IN AN EMBEDDED SYSTEM

What you really need is a small, self-contained, standalone Linux that can be booted from the boot media of your embedded system. The minimum requirement for Linux is a little less than 2MB of disk space, including the boot system, root system, and utilities.

ON THE CD

The *boot system* is essentially the Linux operating system that loads a root file system that contains utilities and files required to run the rest of the operating system. We will discuss the process of creating a minimum Linux system on a floppy disk only. The method is applicable to other file system media such as Flash RAM, NV RAM, and so forth; the only difference is the instructions for copying the file systems to the desired media. It is easier to create a two-floppy system, one for the boot and the other for the root file system. However, you can combine the two into a single floppy if the total size of the file system fits on a 1.44M floppy disk. An example root file system is provided in the companion CD directory /book/chapter1/miniLinux. There are two scripts in the directory, "makeBootDisk" and "makeRootDisk." Please read the readme.txt file provided in the directory of /book/chapter1 for special instructions on how to use the scripts. These scripts can automate the process of creating a boot floppy and a root floppy that you can use to test a minimum Linux implementation that requires a floppy disk system only.

You need to copy the contents of the CD directory from /book/chapter1 to your root file system using the following command:

```
#mount /dev/cdrom -t iso9660 /mnt/cdrom
#mkdir /book/chapter1
#cp -r /mnt/cdrom/book/chapter1/* /book/chapter1/.
```

Building a Boot File System

ON THE CD

Creating a boot disk requires configuring and creating an image of the Linux kernel that is suitable to your specific needs. Configuration of the kernel means adding and removing modules such as CD-ROM, IEEE 1394 Firewire, USB, tape system, and so forth. Obviously, the more options you add during the kernel configuration, the greater the size of the kernel. Therefore, keep it to a minimum and create an image of the kernel that pertains to your specific needs. A default configuration file is provided in the companion CD directory /book/chapter1/miniLinux/miniKernel.config that you can load for a sample startup configuration. The following steps will create a boot image of the Linux kernel on a floppy disk, assuming you are using Red Hat 7.1. Alternately, you can use the script makeBootDisk provided in the companion CD /book/chapter1 to create a boot disk of a minimum Linux system. Please read the readme.txt file of /book/chapter1 for an explanation of the script commands.

1. Enter the following:

```
#cd /usr/src/linux-2.4.2
#make xconfig
```

   The GUI Linux Configuration Kernel window will appear. Click on the "Load Configuration from the file" button and a text box will appear. Enter the filename /book/chapter1/minKernel.config and click OK. Click Save and Exit and you will be returned to the shell prompt.
2. Insert a disk in the floppy and create the boot image with the following command:

```
#make clean
#make bzdisk
```

3. Change the RAMDISK word in the boot floppy so the kernel will be configured to look for the root file system from the floppy media also. The –R option sets the "rootflag," and the –r option with offset 49152 (bit pattern

0xC000) informs the kernel to prompt for loading root disk and load the root file system from sector 0.

```
#rdev /dev/fd0 /dev/fd0
#rdev -R /dev/fd0 0
#rdev -r /dev/fd0 49152
```

This will create the boot floppy disk of your choice. Next, we'll create the root file system floppy disk.

Building a Root File System

Basic requirements for a root file system include:

- The basic file system structure
- Minimum set of directories: /dev, /proc, /bin, /etc, /lib, /usr, /tmp
- Basic set of utilities: sh, ls, cp, mv, etc.
- The programs specific to the embedded applications
- Minimum set of config files: rc, inittab, fstab, etc.
- Devices: /dev/hda, /dev/ttys, /dev/fd0, etc.
- Runtime library to provide basic functions used by utilities

ON THE CD

Creating the root file system involves selecting all the files necessary to perform the desired operation. If you created a separate partition (during the fdisk process) approximately the size of your embedded system boot disk, place all the files necessary to run your system there. Try to execute the code from there; if you are successful, you have a working embedded Linux. You could also use a ramdisk to simulate a disk drive. The Linux loader "Lilo" specifies a 4MB default ramdisk. The following setup uses the default ram disk created during the system startup for the root file system of our minLinux implementation. You could also use the script makeRootDisk provided on the companion CD (directory /book/chapter1) to create a miniLinux root file system on a floppy. Once the Linux is loaded using the boot disk, you will be prompted to insert the root disk for loading the file system and start the operating system.

The following setup creates the root file system that can be placed on a single floppy (insert a floppy prior to executing the commands); alternately, you could use the script makeRootDisk as shown here:

```
#cd /book/chapter1
#./makeRootDisk
```

1. You must clear the contents of the device by filling with zeros; you can use the following command:

```
#dd if=/dev/zero of=/dev/ram0 bs=1k count = 4096
```

2. Create the ext2 file system on the boot media:

```
#mke2fs –m 0 –I 2000 /dev/ram0
```

3. Create a mount point for your device under root file system directory, such as *mnt*:

```
#mkdir /mnt/ram
```

4. Mount the device:

```
#mount /dev/ram0 /mnt/ram
```

The Contents of the File System

Create the following set of directories on the device /mnt/ram with mkdir:

/dev	Place all device files here
/proc	Linux internal use
/etc	System configuration files
/sbin	System binaries
/bin	More binary files
/lib	Runtime shared libraries
/mnt	A mount point for other disks
/usr	Additional utilities and applications

Directory Contents

ON THE CD

Please see the contents of the /book/chapter1/rootfs/ directory on the companion CD as explained in the script makeRootDisk for a listing of the files for a minimum configuration.

/dev

All device-specific files are stored here. The list of device files can be copied from an existing system using the command:

```
cp –dpR /dev /mnt/ram/dev
```

The dp switch ensures that the symbolic links are copied as links. You need to copy only the device files that are essential for the system, but the following files must be included you may remove the rest:

```
console, fd0, kmem, mem, null,  tty0, tty1, tty2, tty3, tty4, tty5,
tty6
```

/etc

This directory contains important configuration files.The rc should have the following contents:

```
#!/bin/sh
/bin/mount -av
/bin/hostname xyz
```

rc.d/ System startup and run-level change scripts.

fstab List of file systems to be mounted as shown here:

```
/dev/ram0        /              ext2    defaults
/dev/fd0         /              ext2    defaults
/proc            /proc          proc    defaults
```

inittab Parameters for the init process, the first process started at boot time. The following is a minimal inittab file:

```
id:2:initdefault:
si::sysinit:/etc/rc
1:2345:respawn:/sbin/mingetty  tty1
2:23:respawn:/sbin/mingetty  tty2
```

passwd Critical list of users, home directories, etc.

group User groups.

termcap The terminal capability database.

shutdown Handle shutdown command.

/bin and /sbin

Copy the basic utilities into this directory, such as:

```
ls, mv, cat, and dd
```

/usr

Include the following programs:

```
init, getty, login, mount, and  a link from sh to the shell
```

/lib

Place all necessary shared libraries and loaders. The following are mandatory:

```
libc library, libc.so.N
```

Name Service Switch (NSS)

If you are using glibc or libc6, you must make provisions for name services or you will not be able to log in. The file /etc/nsswitch.conf controls database lookups for various services.

The following is a simple nsswitch.conf file:

```
passwd:     files
shadow:     files
group:      files
hosts:      files
services:   files
networks:   files
protocols:  files
rpc:        files
ethers:     files
netmasks:   files
bootparams: files
automount:  files
aliases:    files
netgroup:   files
publickey:  files
```

This specifies that only local files provide every service. You will also need to include /lib/libnss_files.so.X, where X is 1 for glibc 2.0, and 2n for glibc 2.1. This library will be loaded dynamically to handle the file lookups.

Modules

Most likely, you will be using a modular kernel. A modular kernel allows dynamic loading of device drivers. Place all the device modules in the directory /lib/modules.

Add insmod, rmmod, lsmod, modprobe, depmod, and swapout to your bin directory. Include /etc/conf.modules.

```
/var/log is for kernel runtime messages.
```

If you had copied the root file system from the accompanied CD directory as described in the boot disk setup then the following commands will create the root system floppy:

```
#ldconfig -r /mnt/ram
#cd /mnt
#umount /mnt/ram
#dd if=/dev/ram0 bs=1k | gzip -v9 > rootfs.gz
#dd if=rootfs.gz of=/dev/fd0 bs=1k
```

SUMMARY

This chapter discussed building your Linux system as a development machine, and the essential parts of creating an embedded Linux system. Although we described the procedure for installing Red Hat Linux, you can use other distributions such as Mandrake, SuSE, and so forth. We also discussed the partitioning of the hard disk with the fdisk utility.

We covered:

- Linux as a development system
- Partitioning of a hard disk using the fdisk utility
- Linux installation using the Red Hat distribution
- Essential components of an embedded Linux

2 The User Interface

In this chapter

- Database Server (PostgreSQL)
- Domain Name System
- HTTP/Web Server (Apache)

INTRODUCTION

The Internet has become an essential part of our daily life; we receive e-mail, read newspapers, chat, and pay our bills on the Internet. However, we seldom see the user interface of an embedded system design on the Internet. In this chapter, we will try to break the mold and explain how simple the Internet technology is, so next time there is an embedded design project, think *Internet* for the user interface component of the design. If you feel uncomfortable with the word *Internet* because it makes your system vulnerable and prone to intrusion from outside, just settle for intranet access. Technically, intranets and the Internet are the same. The only difference is that for the Internet, you have to register your domain name, thereby making your domain name public. If that makes you uneasy, stay with the intranet and use all the bells and whistles the Web interface has to offer.

By the time you finish this chapter (and have followed the necessary instructions), you will have a functional Web hosting server just like a dot com. The basic ingredients you need are Web server, DNS server, and SQL server. These services form the backbone of a company's presence on the Internet, and provide the basis for an organization's Internet and intranet services needs—the basic services we have chosen the Linux operating system for. So far, the Web has proven itself as the best means of displaying information, and we would like to have the embedded system design presented on the network.

Some applications can be designed efficiently with the help of Internet technology only. Consider a flow meter sensor out in the desert receiving data for a pipeline transmission system, or a printer where you store all your documents for archival purposes for later printing. Although the user interface requirements for each are different, they have some things in common: the information for both must be secured, the access must be restricted to authorized users only, and a robust transmission mechanism must be in place for a viable system solution. A complete inhouse software development might appear to be the best solution, but the implementation can be very expensive.

The recommended solution is to implement a database server (SQL or DBM) to log the sensor data. Implement a script or an executable (Perl script or a C compiled program) to siphon the data out of the database. Implement a Web server or a Socket server to channel the data on to the user site, and implement the DNS server or Socket port to identify the specific device in which we are interested. Figure 2.1 depicts the flow of information between the functional blocks.

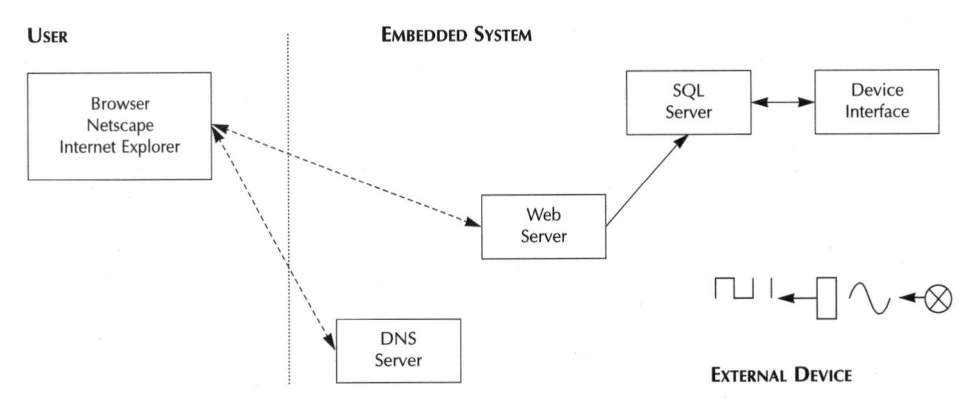

FIGURE 2.1 An embedded system design with user interaction via a browser.

The advantage of using a Web-based solution is that it is platform independent. An application launched from a Linux server can be viewed on a Web browser from any other operating system (such as Windows 9x, Macintosh, or Unix machines), and from as faraway as you wish.

You will be amazed to see how easy and inexpensive it is to implement such a scheme on Linux. Linux is highly scalable. A complete embedded Linux can be ported on as little as 2MB disk space; of course, it will not support all the bells and whistles of SQL server, Web server, and DNS server. There are alternatives for these utilities, which we discuss in Chapter 4, "Tasks and Interprocess Communication," when we develop a socket server and client application programs, and demonstrate a low-cost data acquisition system as an embedded system design.

The following tutorial describes the basic server setup for the three essential services mentioned previously: SQL server, DNS server, and Web server. The setup assumes you have a complete Linux server such as Red Hat 7.1 or other Linux distribution with development software installed on your system. The setup results in a virtual hosting system on Linux similar to an Internet service provider (ISP) where multiple domains are registered. We will use the technique developed here in subsequent sections to implement some real-life embedded design projects—a door lock mechanism with coded key, an employee time and attendance log, and a system for payroll and manufacturing cost management.

The DNS server provides name resolution services to the browser and the Web server. The SQL database server provides the common gateway interface (CGI) scripts and access to the stored information, and the Web server is responsible for channeling the data to and from the user.

DATABASE SERVER (POSTGRESQL)

PostgreSQL is a sophisticated transaction-based database system, capable of supporting extremely large databases at par with systems such as Sybase and Oracle. The fact that it comes free with Linux is one good reason to design your projects around it. As all other servers, PostgreSQL runs continually, waiting for clients to connect. There is no direct interaction between the server and the user; the user must use a client program such as *psql* to communicate with the server. Later in the chapter, we provide a tutorial that uses *psql* commands to set up a database for our embedded system design project "Employee Entrance Time Log and Security System" (we discuss the complete project design in Chapter 9, "Embedded System Design Projects"). PostgreSQL can support several databases simultaneously, but running the server for the first time requires initializing a database system with the *initdb* command. The parameters to initdb are:

```
-D dbdir
```

This option specifies where in the file system the database should be stored.

```
-L libdir
```

initdb needs a few input files to initialize the database. One of the files, global1.bki.source, is traditionally installed along with the others in the library directory (e.g., /usr/local/pgsql/lib).

The –D option and the –L tell the initdb program where to find the libraries.

For security purposes, the PostgreSQL server is not being run as a root user, so the *postmaster* program is being used to start the server at boot time. The consequence of not being a root user is that the X (Windows) server might refuse the connection to the database server. A simple solution to this problem is to use the *xhost* + command that grants X server access to everyone.

The following configuration initializes a database system and creates a database for us to use in our embedded system design projects.

In this configuration:

- The database system will be initialized.
- A database named "timelog" will be created. Our design project "Employee Entrance Time Log and Security System" will use this database.
- The database will be populated with two tables, the "employee" table for the employee name, ID, and so forth, and the "entrancelog" table where timestamps for employee time-in and time-out are stored.
- The interactive program *psql* will be used to create the tables and populate the columns and values of the tables.

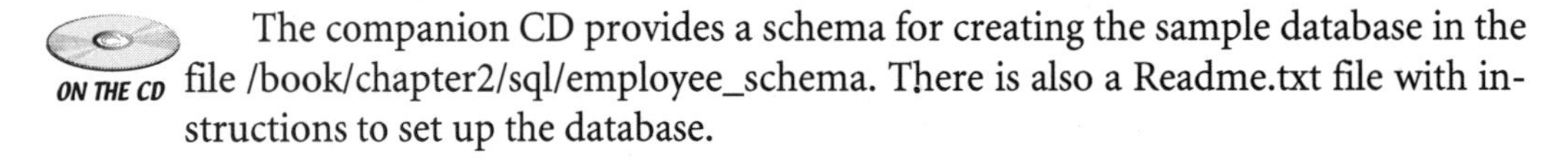

The companion CD provides a schema for creating the sample database in the file /book/chapter2/sql/employee_schema. There is also a Readme.txt file with instructions to set up the database.

PostgreSQL Server Setup

If you selected the PostgreSQL option at installation time, you can proceed with the server setup. Red Hat Linux uses the Red Hat Package Manager (RPM) to install and upgrade software packages, but you can also download the latest software from the official Web site www.postgresql.org.

To download and install the PostgreSQL:

1. Open a command shell and create a directory /sql.

```
# mkdir /sql
# cd /sql
```

2. Download the latest version from www.postgresql.org into the directory /sql, and untar and unzip the file with the following shell command:

```
# tar –xvzf  postgresql-7.3.2.tar.gz
# cd postgresql
```

3. Make and install the PostgreSQL with the following shell command:

```
# make
# make install
# make setup
```

Using Red Hat RPM

The following procedure specifically refers to Red Hat version 7.1. Your installation might be different; if so, simply change the PostgreSQL version number for installation.

1. Insert the Red Hat CD, open a command shell, and mount the CD-ROM.

```
# mount –t iso9660 /dev/cdrom /mnt/cdrom
# cd /mnt/cdrom/RedHat/RPMS
```

2. At the command prompt, execute the dir program as follows to find out the version provided on your CD.

```
# dir  | grep postgresql
```

3. Install the appropriate PostgreSQL software provided on your CD. (The following command sequence is just an example of the command sequence provided on the Red Hat 7.1 CD. Your distribution may have a different version number.) At the command prompt, execute the rpm program as follows (the *** in the following command sequence means the filename should be filled in by pressing the Tab key), and reboot the system.

```
# rpm -i  postgresql-***
# rpm -i  postgresql-server-***
# rpm -i  postgresql-devel-***
# rpm -i  postgresql-perl-***
```

4. Make a link in the rc5.d directory that would start the server at boot time and reboot the machine.

```
# cd /etc/rc.d/rc5.d
#ln –s ../init.d/postgresql S91postgresql
```

Server Setup Procedure

1. Allow PostgreSQL to use the X display.

```
# xhost +
```

2. Change the ownership of /var/lib/pgsql and /usr/lib/pgsql directory to postgres.

```
# chown -R postgres:postgres /var/lib/pgsql
# chown -R postgres:postgres /usr/lib/pgsql
```

3. Change to user postgres.

```
# su postgres
```

4. Initialize the database system using the –L option for the install directory, and the –D option for the database directory.

```
$ initdb –L /usr/lib/pgsql –D /var/lib/pgsql
```

5. Switch back to the root user, and start the database system with the following command at the prompt:

```
$ su
password:  password
# /etc/rc.d/init.d/postgresql restart
```

ON THE CD

6. A database schema is provided in the companion CD file /book/chapter2/sql/employee_schema. Insert the book CD and copy the schema to your local directory. You can proceed to step 7 if you don't have the CD, but you might have to populate the database manually using SQL commands directly.

```
# mkdir  /book/chapter2/sql
# cp /mnt/cdrom/book/chapter2/sql/employee_schema  /book/chapter2/sql/.
```

7. Switch back to the user postgres and create the database timelog.

```
# su postgres
$ createdb timelog
```

8. Start the interactive backend psql and open the database timelog.

```
$ psql timelog
```

9. Populate the database from the schema provided in the file /book/chapter2/sql/employee_schema using the import command of psql. You can proceed to the next section and create the database tables manually for an interactive session with psql if you do not have the access to the book CD.

```
timelog=$ \i /book/chapter2/sql/employee_schema
```

10. Verify the database table entries using SQL queries.

```
timelog=$ SELECT * From table employee;
```

Setting the Time Log Database Schema

In embedded system design, the database plays the role of intermediary between the system and the user. It is also a place where information is persistent and the state of the system is preserved. Just as a flowchart helps design the software algorithm, a database schema is a roadmap to information in the database. The following is the schema for the application "Employee Entrance Time Log and Security System." The application allows an employee to log time-in and time-out through a Web browser such as Netscape, and a supervisor can view an employee's time sheet and upload time records into a file for accounting purposes.

The database schema for the Time and Attendance database contains two tables:

- **Employee**: This table contains employee information, including ID, Last Name, First Name, Middle Initial, Active Status, and the employee's supervisor's ID (Table 2.1).

TABLE 2.1 Database Schema for the Employee Table

ID	Last Name	First Name	MI	Active	Supervisor

TABLE 2.2 Database Schema for the Timelog Table

ID	Time In	Time Out

- **Timelog:** This table contains employees' login information (Table 2.2).

The tables and the columns in the schema must satisfy the following constraints:

1. Employee.ID is the primary key. It is unique and must not be NULL.
2. Employee.LastName must not be NULL.
3. Employee.FirstName must not be NULL.
4. Employee.Status has a default value "Employee."
5. Employee.Supervisor (being himself an employee) is a key that refers back to Employee.ID, and must be allowed to be NULL. It allows us to enter our first entry.
6. TimeLog.ID is a foreign key that refers to Employee.ID as the primary key.

Creating the Employee and Timelog Database Tables

1. Start a command shell, and switch user to the postgres.

```
#   su postgres
```

2. Create the database "timelog."

```
$ createdb timelog
```

3. Launch the backend program psql to connect to the database.

```
$ psql -s timelog
Welcome to psql, the PostgreSQL terminal
type \? for help on slash commands
type \q to quit
type \h for help
type \g or terminate with semicolon to execute query
You are currently connected to the database: postgres
```

4. Create the Employee table at the pgsql command prompt.

```
timelog=#  create table employee
timelog-> (ID char (10) NOT NULL UNIQUE PRIMARY KEY,
timelog-> password char (10) NOT NULL,
timelog-> LastName char(16) NOT NULL,
timelog-> FirstName char(16) NOT NULL,
timelog-> MI char(2),
timelog-> Active (2) DEFAULT  'A',
timelog-> Supervisor  char(10));
```

5. Populate the Employee table with the administrator name and ID at the pgsql command prompt.

```
timelog=#  insert into
timelog-> employee (ID,password,LastName,FirstName)
timelog-> values ('0000','password', 'Admin', 'Root');
```

6. Create the Entrancelog table.

```
timelog=# create table entrancelog
timelog=> (ID char (10), timein timestamp DEFAULT CURRENT_TIMESTAMP,
timelog=> timeout timestamp DEFAULT CURRENT_TIMESTAMP);
```

7. Grant access to PUBLIC so the database can be used by the Web server.

```
security=> GRANT ALL on entrancelog to nobody;
security=> GRANT ALL on employee to nobody;
```

8. You can view all the table entries with the SELECT command as follows:

```
timelog=> select * from employee;
```

9. Enter \q to exit the psql.

```
security=> \q
```

The database is now ready for application interface. Next is the Domain Name System (DNS) server setup.

The timelog database is used in the Web application project "Employee Entrance Time Log System" as described in the next section. The Web application is based on the Perl script that interfaces with the database upon request from the

Web server. The application requires the server setup of DNS and the Apache Web server.

THE DOMAIN NAME SYSTEM

The Domain Name System, or DNS, is the backbone of the Internet Naming Service to which we have become so accustomed. Every time you request a Web page on your browser, DNS gets involved. In order to understand the function of DNS, you need to understand the naming system on the Internet. The Internet is divided into *domains* or *zones* (we will use *domain* and *zones* interchangeably; they are the same), and each domain has a specified number of computers. The domains are called *networks*, and the computers are called *hosts*. TCP/IP is the defacto standard for the communication protocol on the Internet, and as a rule, every computer on the Internet must be associated with a unique identification number, the IP address. The IP address is a 32-bit number, created by combining the network number and the host number. However, because remembering numbers can be difficult, we assign names to our computers as well.

It is the DNS server's responsibility to resolve IP addresses of the computer names that belong to a specific domain. An application such as Netscape Navigator consults a DNS server first and obtains the IP address of the requested host when a user enters the name of a Web site. For example, when you enter www.company.com in the browser address box, you are essentially requesting the access of the computer "www" in a domain "company.com." DNS provides, upon request, the specific IP address of the computer in the specified domain. DNS can be configured to forward the request to another DNS server if it cannot resolve the name request on its own. However, it is smart enough to remember the name and stores the information in its cache file, so the next time a request comes in for the same site, it can return the information right away.

In our scheme of embedded system design projects, we use the following fictitious domains and hosts to provide the users access to the system information.

- www.project.com
- www.printer.com
- www.digital.com
- www.timelog.com

If you are planning to use the Apache Web server, the DNS server setup is mandatory. However, not all embedded designs can accommodate a full-blown Web server. We develop a standalone Web server in Chapter 4, "Tasks and Interprocess Communication," that is ideally suited for systems with space constraints.

For now, we will assume that our product will be presented on the intranet using the Apache Web server. The domains we have selected reside on the same machine; thus, any request from the browser (such as Netscape) points to the same IP address at which the server resides. The Web server is designed to redirect the request to a specific directory in the system, based on the domain name.

The DNS server is basically a database of host information, and the data are available via DNS Server (called *name server*) through a client-server mechanism. The *resolver* is a client that submits queries to the server and returns the host information. DNS is not the only method to retrieve host information; the */etc/hosts* file can also be set up to provide host information. In fact, you must provide one host file with at least one entry for the primary host information where the DNS resides, so the information regarding the primary host is available to the system immediately.

Installing DNS Using Red Hat RPM

If you have already checked the option of install DNS at Linux installation time then you may skip this procedure and go directly to the section *DNS Server Components* and start setting the domains and zones. This section describes the procedure of installing DNS server from Red Hat 7.1 distribution. Simply change the Apache version number if you are using different distribution.

1. Insert the Red Hat CD, open a command shell, and mount the CD-ROM.

```
# mount –t iso9660 /dev/cdrom /mnt/cdrom
# cd /mnt/cdrom/RedHat/RPMS
```

2. At the command prompt, execute the dir program as follows to find out the version provided on your CD.

```
# dir  | grep bind
```

3. Install the appropriate bind software provided on your CD. (The *** in the following command sequence means the file name should be filled-in by pressing tab key). Execute the rpm program as follows, and reboot the system.

```
# rpm -i  bind***
# rpm -i  bindconf***
# rpm -i  bind-util***
```

4. Start the bind server.

```
# cd /etc/rc.d/init.d
#./named start
```

5. You may create a link to the startup script in the rc5.d directory for the service to start automatically at boot time.

```
# cd /etc/rc.d/rc5.d
#ln –s ../init.d/named S92named
```

The DNS Server Components

The DNS server can be designated as an authoritative server over any zone or domain, even if the DNS server itself resides in that zone. Thus, one name server can provide naming information for multiple zones as if it has authority over that zone. The Linux DNS server is *bind* (Berkley Internet Name Domain), and the configuration file for the bind *is /etc/named.conf*. The DNS server has the following three functions:

- Forward name resolution (name-to-IP-address translation)
- Reverse name lookup (IP-address-to-name translation)
- Cache information (stores the host's information not found in the primary zone)

A default configuration file /etc/named.conf is created when the DNS is installed. Listing 2.1 shows the contents of the file. The file is divided into the option section and three zone information sections. The option section (lines 2 through 12) line 3 indicates the directory path (/var/named) where the DNS zone information files are being stored. The three zone information sections are the caching zone (lines 16 through 19), the localhost zone (lines 20 though 24) and reverse naming information (lines 25 through 29). The most important information in the zone section is the name of the file where all individual host information for the zone is being stored.

Adding a new zone for the DNS requires:

- Adding a zone information section in the named.conf file
- Creating the host information file pertaining to the specified zone

If the hosts in a DNS server zone have physical IP addresses, then it is necessary to add a reverse name lookup section in the named.conf file. In which a case the reverse name resolution file must be created in the directory path specified in the option section of the /etc/named.conf file. In our scheme of virtual Web hosting setup, all the hostnames in the fictitious domains are canonical names; thus, a reverse name lookup service is unneeded.

- Creating the reverse name resolution file in the directory path specified in the option section of the /etc/named.conf file.

The reverse name resolution zone section starts with the network IP number plus in-addr.arpa; for example, 0.168.192.in-addr.arpa (omitting the host section from the IP number). Listing 2.1 shows the contents of the configuration file /etc/named.conf.

LISTING 2.1 The default configuration file /etc/named.conf, created at DNS installation time.

```
1) // generated by named-bootconf.pl
2) options {
3)      directory "/var/named";
4) /*
5)    If there is a firewall between you and nameservers you want
6)    to talk to, you might need to uncomment the query-source
7)    directive below.  Previous versions of BIND always asked
8)    questions using port 53, but BIND 8.1 uses an unprivileged
9)    port by default.
10) */
11) // query-source address * port 53;
12) };
13) //
14) // a caching only nameserver config
15) //
16) zone "." IN {
17)    type hint;
18)    file "named.ca";
19) };
20) zone "localhost" IN {
21)    type master;
22)    file "localhost.zone";
23)    allow-update { none; };
24) };
25) zone "0.0.127.in-addr.arpa" IN {
26)    type master;
27)    file "named.local";
28)    allow-update { none; };
29) };
```

Authority over a Zone or Domain

The /etc/named.conf file carries only zone information. The individual host information is presented in a separate file as specified by the zone information. Adding a new zone or domain is a two-step process. First, add the new zone information to the named.conf file. Second, create a new file for all the host information residing in the domain.

ON THE CD

Next, we discuss the procedure for adding four domains (project.com, timelog.com, digital.com, and printer.com) for which our DNS server will be an authoritative name server. The setup requires editing the named.conf file, but you can copy the file from the companion CD using the shell command.

1. Mount the CD-ROM.

```
# mount –t iso9660 /dev/cdrom /mnt/cdrom
```

2. Copy the *host* file.

```
# cp /mnt/cdrom/book/chapter2/dns/hosts /etc/
```

3. Copy the *resolv.conf* file.

```
# cp /mnt/cdrom/book/chapter2/dns/resolv.conf /etc/
```

4. Copy the *network* file.

```
# cp /mnt/cdrom/book/chapter2/dns/network /etc/sysconfig/
```

5. Copy the *ifcfg-eth0* file.

```
# cp /mnt/cdrom/book/chapter2/dns/network-scripts/ifcfg-eth0
/etc/sysconfig/network-scripts/
```

6. Copy the *named.conf* file.

```
# cp /mnt/cdrom/book/chapter2/dns/named.conf /etc/
```

The following setup is for DNS zone information files:

1. Copy the *digital.com* file.

```
# cp /mnt/cdrom/book/chapter2/dns/digital.com /var/named/
```

2. Copy the *analog.comt* file.

```
# cp /mnt/cdrom/book/chapter2/dns/analog.com  /var/named/
```

3. Copy the *project.comt* file.

```
# cp /mnt/cdrom/book/chapter2/dns/project.com /var/named/
```

4. Copy the *timelog.com* file.

```
# cp /mnt/cdrom/book/chapter2/dns/timelog.com /var/named/
```

5. Copy the *printer.comt* file.

```
# cp /mnt/cdrom/book/chapter2/dns/printer.com /var/named/
```

6. Copy the *localhost.zone* file.

```
# cp /mnt/cdrom/book/chapter2/dns/localhost.zone /var/named/
```

7. Copy the *named.local* file.

```
# cp /mnt/cdrom/book/chapter2/dns/named.local /var/named/
```

8. Copy the *named.ca* file.

```
# cp /mnt/cdrom/book/chapter2/dns/named.ca /var/named/
```

Adding the project.com Domain

Append the following text to the /etc/named.conf file using your favorite text editor. Line 1 is the zone name "project.com," line 2 specifies the primary server for the domain, and line 3 is the host information filename being created in the /var/named directory ("project.com").

```
1)  zone "project.com" IN {
2)     type master;
3)     file "project.com";
4)     notify no;
5)  };
```

Adding the printer.com Domain

Similar to the project.com, append the following text to the /etc/named.conf file. The zone name is "printer.com," and the host information file is "printer.com."

```
1) zone "printer.com" IN {
2) type master;
3) file "printer.com";
4) notify no;
5) };
```

Adding the digital.com Domain

Append the following text to the /etc/named.conf file. The zone name is "digital.com," and the host information file is "digital.com."

```
1) zone "digital.com" IN {
2) type master;
3) file "digital.com";
4) notify no;
5) };
```

Adding the timelog.com Domain

Append the following text to the /etc/named.conf file. The zone name is "timelog.com," and the host information file is "timelog.com."

```
1) zone "timelog.com" IN {
2) type master;
3) file "timelog.com";
4) notify no;
5) };
```

Host Information Setup

The completed configuration file /etc/named.conf is presented in Listing 2.2, and shows the four new zones: project.com, printer.com, digital.com, and timelog.com. However, the information is still incomplete. We need to create the actual host files mentioned in each zone information section.

LISTING 2.2 The /etc/named.conf with the four additional zones: project.com, printer.com, digital.com, and timelog.com.

```
1) // generated by named-bootconf.pl
2) options {
```

```
3)      directory "/var/named";
4)  /*
5)     If there is a firewall between you and nameservers you want
6)     to talk to, you might need to uncomment the query-source
7)     directive below.  Previous versions of BIND always asked
8)     questions using port 53, but BIND 8.1 uses an unprivileged
9)     port by default.
10) */
11) // query-source address * port 53;
12) };
13) //
14) // a caching only nameserver config
15) //
16) zone "." IN {
17)    type hint;
18)    file "named.ca";
19) };
20) zone "localhost" IN {
21)    type master;
22)    file "localhost.zone";
23)    allow-update { none; };
24) };
25) zone "0.0.127.in-addr.arpa" IN {
26)    type master;
27)    file "named.local";
28)          allow-update { none; };
29)       };
30) zone "project.com" IN {
31)    type master;
32)    file "project.com";
33)    notify no;
34) };
35) zone "printer.com" IN {
36)    type master;
37)    file "printer.com";
38)    notify no;
39) };
40) zone "digital.com" IN {
41)    type master;
42)    file "digital.com";
43)    notify no;
44) };
45) zone "timelog.com" IN {
46)    type master;
```

```
47)    file "timelog.com";
48)    notify no;
49) };
```

Once a domain is added to the name server authority, individual host's information can be added as needed. Presently, the only host we will be adding is the www host. The named.conf file's option section specifies the directory in which the host information files should be located (line 3 of Listing 2.2 showing /var/log/named directory). The four host files (project.com, digital.com, printer.com, and timelog.com) need to be created in the specified directory. Listing 2.3 shows the contents of a typical host file.

LISTING 2.3 The host file for the zone project.com.

```
1)  @        IN    SOA    ns.project.com.    hostmaster.ns.project.com. (
2)      2001052401 ; serial
3)      3600 ; refresh
4)      900 ; retry
5)      1209600 ; expire
6)      43200 ; default_ttl
7)  )
8)  @        IN    MX     5                  ns.project.com.
9)  @        IN    NS     ns.project.com.
10) www      IN    CNAME                     ns.project.com.
11) ns       IN    A      192.168.0.40
```

The following is an explanation of the information in each line of the zone information file.

SOA Record

The first entry in each of these files is the *start of authority* SOA () resource record. The @ in column 1 represents the current zone that is being looked at (since we got here by first reading the zone "project.com" information in the named.conf file); the @ is replaced with "project.com." IN is the Internet. SOA is the record type. The "ns.project.com" after SOA () is the primary master name server for this data. The hostmastrer.ns.project.com. is the mail address of the person in charge of the data. The parenthesis following the mail recipient spans multiple lines, and the contents provide information regarding the serial number, the refresh rate, the retry count, the expiration time, and default time-to-live (TTL).

NS Record

The lines containing "NS" (name server) are the name server resource records; for example:

```
@           IN      NS      ns.project.com.
```

Again, the @ represents the zone being looked at (project.com), and the name after NS (ns.project.com) specifies the name server for the zone project.com.

Address Records

The lines with A (address) map names to address; for example:

```
ns          IN      A       192.168.0.40
```

This is the line where you add new computer names and associate them with the IP address. The name in the first column is the assigned computer name, followed by IN for Internet, A for address, and then the IP address assigned.

Alias Records

The lines containing CNAME (canonical name) map alternate names to an address; for example:

```
www         IN      CNAME   ns.project.com.
```

If you are assigning more than one name to the same computer, this is the line. The "www" is the nickname for the computer ns.project.com.

PTR Records

The lines containing PTR (pointer) provide address to name mapping; for example:

```
40.0.168.192.in-addr.arap.      IN      PTR     ns.project.com.
```

These lines provide reverse name lookup capability to DNS.

MX Records

The lines containing MX (Mail Exchanger) provide mail routing information; for example:

```
@           IN      MX      5       ns.project.com.
```

The record specifies the mail exchanger for the domain name. If there is more than one MX record, the line specifies the degree of preference for the mail exchanger for the domain. The number followed by MX is the relative preference assigned to each mail server, so the mailer should try the preferred one (higher number) first, and then try the alternate server in case of failure.

Adding the Virtual Host "www.timelog.com"

ON THE CD

Create the file "timelog.com" in the directory (/var/named) as shown in Listing 2.4, or copy the file from the companion CD directory /book/chapter2/dus/timelog.com.

LISTING 2.4 timelog.com host information file for the domain timelog.com.

```
1) @         IN   SOA    ns.project.com.  hostmaster.ns.project.com. (
2)     2001052401 ; serial
3)     3600 ; refresh
4)     900 ; retry
5)     1209600 ; expire
6)     43200 ; default_ttl
7) )
8) @         IN   MX     5                ns.project.com.
9) @         IN   NS     ns.project.com.
10) www      IN   CNAME                   ns.project.com.
```

Adding the Virtual Host "www.project.com"

ON THE CD

Create the file "project.com" in the directory (/var/named) as shown in Listing 2.5, or copy the file from the companion CD directory /book/chapter2/dus/project.com.

LISTING 2.5 timelog.com host information file for the domain project.com

```
1) @         IN   SOA    ns.project.com.  hostmaster.ns.project.com. (
2) 2001052401 ; serial
3) 3600 ; refresh
4) 900 ; retry
5) 1209600 ; expire
6) 43200 ; default_ttl
7) )
8) @         IN   MX     5                ns.project.com.
9) @         IN   NS     ns.project.com.
10) www      IN   CNAME                   ns.project.com.
11) ns       IN   A      192.168.0.40
```

Note that there are two computers in the domain project.com: "ns" and "www." The "ns" is the primary name being assigned to the computer.

Adding the Virtual Host "www.printer.com"

ON THE CD

Create the file "printer.com" in the directory (/var/named) as shown in Listing 2.6, or copy the file from the companion CD directory /book/chapter2/dus/printer.com.

LISTING 2.6 timelog.com host information file for the domain printer.com.

```
1) @        IN  SOA   ns.project.com.  hostmaster.ns.project.com. (
2) 2001050201 ; serial
3) 3600 ; refresh
4) 900 ; retry
5) 1209600 ; expire
6) 43200 ; default_ttl
7) )
8) @        IN  MX    5                ns.project.com.
9) @        IN  NS    ns.project.com.
10) www     IN  CNAME                  ns.project.com.
```

Adding the Virtual Host "www.digital.com"

ON THE CD

Create the file "printer.com" in the directory (/var/named) as shown in Listing 2.7, or copy the file from the companion CD directory /book/chapter2/dus/digital.com.

LISTING 2.7 timelog.com host information file for the domain digital.com.

```
1) @        IN  SOA   ns.project.com.  hostmaster.ns.project.com. (
2) 2001050201 ; serial
3) 3600 ; refresh
4) 900 ; retry
5) 1209600 ; expire
6) 43200 ; default_ttl
7) )
8) @        IN  MX    5                ns.project.com.
9) @        IN  NS    ns.project.com.
10) www     IN  CNAME                  ns.project.com.
```

The PING Command

The fastest way to verify a connection is to use the PING utility. If the requested host is out there somewhere, the PING response will show that the host is present and responding. For example, the following PING command at the shell prompt verifies the server ns.project.com:

```
# ping ns.project.com
    PING ns.project.com (192.168.0.40): 56 data bytes
    64 bytes from 192.168.0.40: icmp_seq=0 ttl=255 time=0.2 ms
    64 bytes from 192.168.0.40: icmp_seq=1 ttl=255 time=0.1 ms
    64 bytes from 192.168.0.40: icmp_seq=2 ttl=255 time=0.1 ms
    —- ns.project.com ping statistics —-
    3 packets transmitted, 3 packets received, 0% packet loss
    round-trip min/avg/max = 0.1/0.1/0.2 ms
```

Receiving a time-out error can mean that the basic host name setup in the DNS is incorrect, or you do not have networking enabled, or you do not have the correct host.conf file. Consult the section *Verifying DNS Functionality* for troubleshooting guidelines.

Verifying the Virtual Host www.timelog.com

Enter the following PING command at the shell prompt to verify that the virtual host setup is correct and that PING returns a valid response:

```
# ping www.timelog.com
    PING ns.project.com (192.168.0.40): 56 data bytes
    64 bytes from 192.168.0.40: icmp_seq=0 ttl=255 time=0.1 ms
    64 bytes from 192.168.0.40: icmp_seq=1 ttl=255 time=0.1 ms
    —- ns.project.com ping statistics —-
    2 packets transmitted, 2 packets received, 0% packet loss
    round-trip min/avg/max = 0.1/0.1/0.1 ms
```

If a valid response is not received or a time-out error occurs, fix your /etc/named.conf or /var/named/timelog.com file.

Verifying the Virtual Hosts

Similar to www.timelog.com, verify the host www.digital.com, www.printer.com, and www.project.com setup of your DNS with the following commands:

```
# ping www.project.com
# ping www.timelog.com
# ping www.analog.com
```

Verifying DNS Functionality

Proper functioning of the DNS facility requires the basic networking of the system to be up and running smoothly. If your PING of a domain does not respond with 0 error, the following is a guideline to help you debug the configuration files.

/etc/sysconfig/network

In order for your system to take part in networking, the environment variable "NETWORKING" has to be set to "yes," and the HOSTNAME specified as shown here. This is done at system startup time by reading the file /etc/sysconfig/network.

Verify that networking is enabled in the /etc/sysconfig/network file with the following shell command and response:

```
# cat /etc/sysconfig/network
    NETWORKING=yesFORWARD_IPV4="yes"
    HOSTNAME="ns.project.com"
    GATEWAYDEV="eth0"
    GATEWAY="192.168.0.40"
```

/etc/sysconfig/network-scripts/ifcfg-eth0

The Ethernet connection requires an IP address, subnet mask, broadcast address, network address, and the device name for your Ethernet connection. A typical ifcfg-eth0file contains the following:

```
# cat /etc/sysconfig/network
    DEVICE="eth0"
    NETMASK="255.255.255.0"
    NETWORK="192.168.0.0"
    IPADDR="192.168.0.40"
    BROADCAST="192.168.0.255"
    ONBOOT="yes"
    BOOTPROTO="none"
```

/etc/hosts

The primary host where the DNS resides must have the file /etc/hosts with at least the entry for the localhost, and the primary host name and the IP address.

Verify the contents of the /etc/hosts file as follows:

```
# cat /etc/hosts
    127.0.0.1       localhost       localhost.localdomain
    192.168.0.40    ns.project.com
```

/etc/host.conf

The *resolver* is the client for the DNS server for resolving name-to-IP translation. The configuration file /etc/host.confbinds as part of the configuration as follows:

```
# cat /etc/hosts
    order hosts,bind
    multi on
```

/etc/named.conf

The configuration file for the bindis /etc/named.conf. See Listing 2.2 for proper configuration of the file.

```
/var/named/timelog.com
/var/named/project.com
/var/named/digital.com
/var/named/timelog.com
```

For each zone information in /etc/named.conf, there is corresponding file in the /var/named directory. Verify the contents as specified in Listings 2.5, 2.6, and 2.7.

HTTP/WEB SERVER

It is hard to imagine a company that does not have a Web page on the Internet. This is a revolution, not an evolution, and embedded system design engineers should take advantage of the opportunity and start designing their systems around Web technology. They have the Linux operating system at their disposal to build and implement the Web, and the Linux tools have proven themselves superior to all others currently available. We have already seen the database server (PostgreSQL) and the DNS server (bind). Now, we will discuss the implementation of the Web server, also known as the HTTP server (Apache), to complete our Internet setup on Linux.

We use the Internet to display "documents"; sound, movies, and the text we see in our browsers are all documents. We receive Web pages in response to a request for a Web site. Web pages are essentially formatted text pages that Web browsers, such as Netscape Navigator, can interpret and display on the screen. The Web server is responsible for responding to the request from the browser by returning the requested document to the user. The forms you see on the Internet are special Web pages that return the values the user has filled in. This is an extremely powerful mechanism for user interface, and the tools and technology available can help you develop Web pages in a very short period of time.

In the past, the user interface for any embedded system design took a considerable amount of a developer's time, but now you can leverage the Web technology to your advantage. You can separate the user interface requirement from the rest of the system, and design dynamic Web pages that give the user the look and feel of system interface.

Perl and Java interpreters provide the database interface to the Web pages, which is our main criteria for using Web technology. The system modules on one side interface with the database and update information that the user requires. The Web server on the other side can launch the Web pages that interface with the database and provide the information to the user. The user interface can be enhanced by the concept of *virtual hosting*, where several different domains are created on the same system and are supported by a single Web server. Each virtual host is independently accessed as if it is a separate machine. The setup we describe next provides virtual-hosting capability on a single Web server for the hosts www.timelog.com, www.project.com, www.printer.com, and www.digital.com. Figure 2.2 illustrates the flow of traffic.

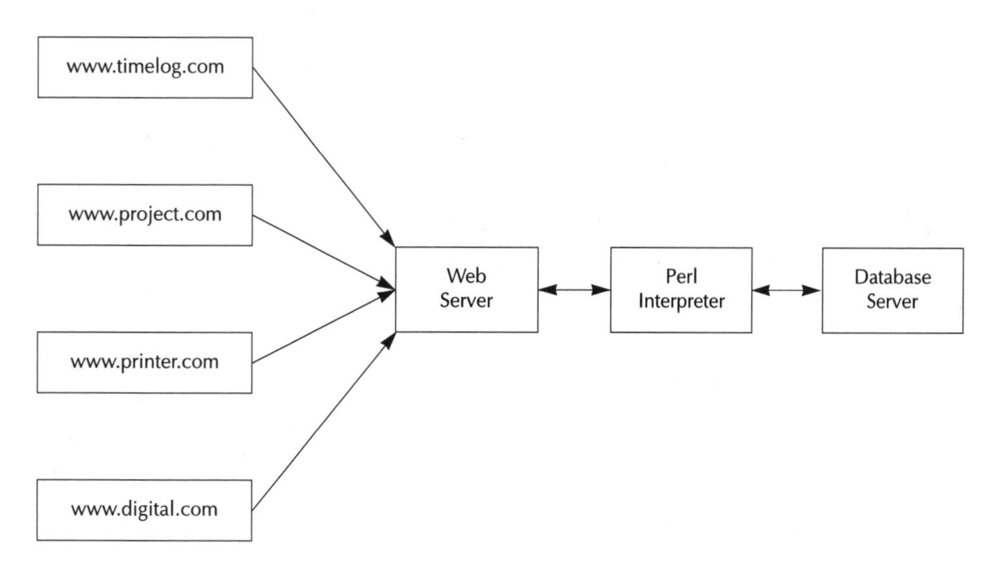

FIGURE 2.2 The flow of information between SQL and the Web server via CGI scripts.

Apache, the HTTP Server

If you cannot find the Apache server in your distribution CD, or if you want to download the latest Apache software, visit the official Web site at www.apache.org. The httpd is the daemon running as the Web server in Linux. It can be a standalone server responding to the TCP/IP port 80 as the default configuration, or the inetd daemon can be configured to launch the httpd in response to a query on port 80. The configuration file for Apache is /etc/httpd/conf/httpd.conf, and has extensive configuration options, including Setting the Document Root, Configuring Directory Options, Allowing Server-Side Includes, Enabling CGI in the Document Root, Enabling CGI as a File Type, to name a few. These will be explained when we discuss the virtual-hosting setup in the next section. The default setup of Apache is just fine for our purposes; the only thing we will be adding is the virtual host's setup specific to our needs. We also need to create the home directories for the virtual hosts, and provide at least a welcome index.html page in each directory so we can verify the setup from the intranet.

Creating the Directories for www.project.com

Use the following shell command to create the default home directory and the default cgi-bin directory for the virtual host www.project.com:

```
# mkdir /home/httpd/html/www
# mkdir /home/httpd/html/www/project
# mkdir /home/httpd/html/www/project/cgi-bin
```

Creating the index.html Page for www.project.com

The home page of every domain should at least provide a welcome banner when a user logs in to the domain. This is by default the index.html file. Create an HTML file as defined in Listing 2.8 using your favorite text editor, and save it as index.html in the directory /home/httpd/html/www/project/. Alternatively, you can copy the file from the companion CD directory using the following commands,

1. Mount the CD-ROM.

```
# mount –t iso9660 /dev/cdrom /mnt/cdrom
# cd /mnt/cdrom/book/chapter2
```

2. Copy the *index.html* file.

```
# cp ./web/project/index.html /home/httpd/html/www/project/.
```

LISTING 2.8 The Welcome index.html file for the domain project.com.

```
1) <HTML>
2) <TITLE> Testing </TITLE>
3) <BODY>
4) <h1>Welcome to timelog.Com</h1>
5) </BODY>
6) </HTML>
```

3. Make the index.html file global read and executable.

```
# chmod 766  /home/httpd/html/www/project/index.html
```

Creating the Directories for www.timelog.com

Use the following shell command to create the default home directory and the default cgi-bin directory for the virtual host www.timelog.com:

```
# mkdir /home/httpd/html/www/timelog
# mkdir /home/httpd/html/www/timelog/cgi-bin
```

Creating the "index.html" Page for www.timelog.com

ON THE CD

The Welcome page for the timelog.com domain is essentially a Web form that requests login information (name and ID) from the user. Notice that the Submit button has a link to the Perl script http://www.timelog.com/cgi-bin/verify_employee.pl. The script is executable and interpreted by the Perl program. This is how we create dynamic Web pages. We discuss these scripts when we develop the application program "Employee Entrance and Time Log" in Chapter 9, "Embedded System Design Projects." Create the Welcome page index.html as defined in Listing 2.9, and save it in the directory /home/httpd/html/www/project/. Alternatively, copy the file from the companion CD directory /book/chapter2/web/timelog/index.html.

1. Copy the *index.html* file.

```
# cp ./web/timelog/index.html /home/httpd/html/www/timelog/.
```

LISTING 2.9 The Welcome index.html file for the domain timelog.com.

```
1)  <HTML>
2)  <TITLE> Testing </TITLE>
3)  <BODY>
4)  <h1>Welcome to Timelog.Com</h1>
5)  Please Enter Name and Password
```

```
6)   <FORM ACTION= "http://www.project.com/cgi-bin/test.pl" method="post">
7)   Name: <INPUT NAME="name" TYPE=text SIZE=10>
8)   Password:<INPUT NAME="password" TYPE=password SIZE=10><BR>
9)   <INPUT TYPE=submit value="Enter">
10)  <INPUT TYPE=submit value="Cancel">
11)  </FORM>
12)  </BODY>
13)  </HTML>
```

2. Make the index.html file global read and executable.

```
# chmod 766 /home/httpd/html/www/timelog/index.html
```

Creating Directories for www.printer.com

Use the following shell command to create the default home directory and the default cgi-bin directory for the virtual host www.printer.com:

```
# mkdir /home/httpd/html/www/printer
# mkdir /home/httpd/html/www/printer/cgi-bin
```

Creating the index.html Page for www.printer.com

ON THE CD

Similar to the Welcome page for the timelog.com domain, create an HTML file as defined in Listing 2.10m and save it as index.html in the directory /home/httpd/html/www/project/. Alternatively, copy the file from the companion CD directory /book/chapter2/web/printer/index.html.

1. Copy the *index.html* file.

```
# cp ./web/printer/index.html /home/httpd/html/www/printer/.
```

LISTING 2.10 The Welcome index.html file for the domain printer.com.

```
1) <HTML>
2) <TITLE> Testing </TITLE>
3) <BODY>
4) <h1>Welcome to Printer.Com</h1>
5) </BODY>
6) </HTML>
```

2. Make the index.html file global read and executable.

```
# chmod 766 /home/httpd/html/www/printer/index.html
```

Creating Directories for www.digital.com

Use the following shell command to create the default home directory and the default cgi-bin directory for the virtual host www.digital.com.

```
#  mkdir /home/httpd/html/www/digital
#  mkdir /home/httpd/html/www/digital/cgi-bin
```

Creating the index.html Page for www.digital.com

Similar to the Welcome page for the timelog.com domain, create an HTML file as defined in Listing 2.11, and save it as index.html in the directory /home/httpd/html/www/project/. Alternatively, copy the file from the companion CD directory /book/chapter2/web/digital/index.html.

1. Copy the *index.html* file.

```
# cp ./web/digital/index.html /home/httpd/html/www/digital/.
```

LISTING 2.11 The Welcome index.html file for the domain digital.com.

```
1) <HTML>
2) <TITLE> Testing </TITLE>
3) <BODY>
4) <h1>Welcome to Digital.Com</h1>
5) </BODY>
6) </HTML>
```

2. Make the index.html file global read and executable.

```
#  chmod 766  /home/httpd/html/www/digital/index.html
```

Virtual Hosts Configuration on Apache

ON THE CD

You might have noticed in our DNS setup that we used several canonical names (CNAME) for the same machine. This is akin to assigning several nicknames to one computer using only one IP address, which is what virtual hosting does for us. With several nicknames—such as www.timelog.com, www.project.com, www.digital.com, and www.printer.com—we can provide Web hosting through one machine and one IP address. It is the Web server's responsibility to direct the traffic to the appropriate domain. Listing 2.12 is the last part of the httpd.conf file that has been modified to provide the virtual-hosting setup. Append the text to the end of the file /etc/httpd/conf/httpd.conf as shown in Listing 2.12, or copy the file from the companion CD directory /book/chapter2/web/httpd.conf.

1. Mount the CD-ROM.

```
# mount -t iso9660 /dev/cdrom /mnt/cdrom
# cd /mnt/cdrom/book/chapter2
```

2. Copy the *httpd.conf* file.

```
# cp ./web/httpd.conf /etc/httpd/conf/
```

LISTING 2.12 The virtual hosting configuration of the httpd.conf file.

```
1)  NameVirtualHost 192.168.0.40
2)  <VirtualHost 192.168.0.40>
3)     ServerName www.timelog.com
4)     DocumentRoot /home/httpd/html/www/timelog
5)     ScriptAlias /cgi-bin/ /home/httpd/html/www/timelog/cgi-bin/
6)     Options ExecCgi Indexes
7)  </VirtualHost>
8)  <VirtualHost 192.168.0.40>
9)     ServerName www.digital.com
10)    DocumentRoot /home/httpd/html/www/digital
11)    ScriptAlias /cgi-bin/ /home/httpd/html/www/digital/cgi-bin/
12)    Options ExecCgi Indexes
13) </VirtualHost>
14) <VirtualHost 192.168.0.40>
15)    ServerName www.printer.com
16)    DocumentRoot /home/httpd/html/www/analog
17)    ScriptAlias /cgi-bin/ /home/httpd/html/www/printer/cgi-bin/
18)    Options ExecCgi Indexes
19) </VirtualHost>
20) <VirtualHost 192.168.0.40>
21)    ServerName www.project.com
22)    DocumentRoot /home/httpd/html/www/project
23)    ScriptAlias /cgi-bin/ /home/httpd/html/www/project/cgi-bin/
24)    Options ExecCgi Indexes
25) </VirtualHost>
```

Each host setup is similar in nature, except for line 1, the NameVirtualHost. That indicates the IP address from where the hosting is being performed. Other options are as follows:

ServerName: Lines 3, 9, 15, 21; identifies the domain.

ServerName www.timelog.com

DocumentRoot: Lines 4, 10, 16, 22; user cannot access anything above this directory.

DocumentRoot /home/httpd/html/www/timelog

ScriptAlias: Lines 5, 11, 17, 23; the file in this directory can be executed or interpreted.

ScriptAlias /cgi-bin/ /home/httpd/html/www/timelog/cgi-bin/

Options: Lines 6, 12, 18, 24;

ExecCgi; CGI scripts are being executed from the cgi-bin directory.

Verifying the Web Server Setup

The following "process status" shell command verifies that the httpd daemon running:

```
[root@ns /root]# ps -ea | grep http
```

If the httpd process is not running, verify that the startup script for the httpd is present in the specified run level. For example, if the run level is 5, the directory /etc/rc.d/rc5.d/ should have an entry for the httpd startup script with the name Sxxhttpd.

```
# cd /etc/rc.d/rc5.d
# ls -al | grep httpd
```

If you do not see a script, one can be created to run httpd at startup time with the following shell command to start and reboot the machine:

```
#cd /etc/rc.d/rc5.d
#ln -s ../init.d/httpd S93httpd
```

Enter the following shell command to verify the Web page for the domain timelog.com as shown in Figure 2.3:

```
# netscape http://www.timelog.com
```

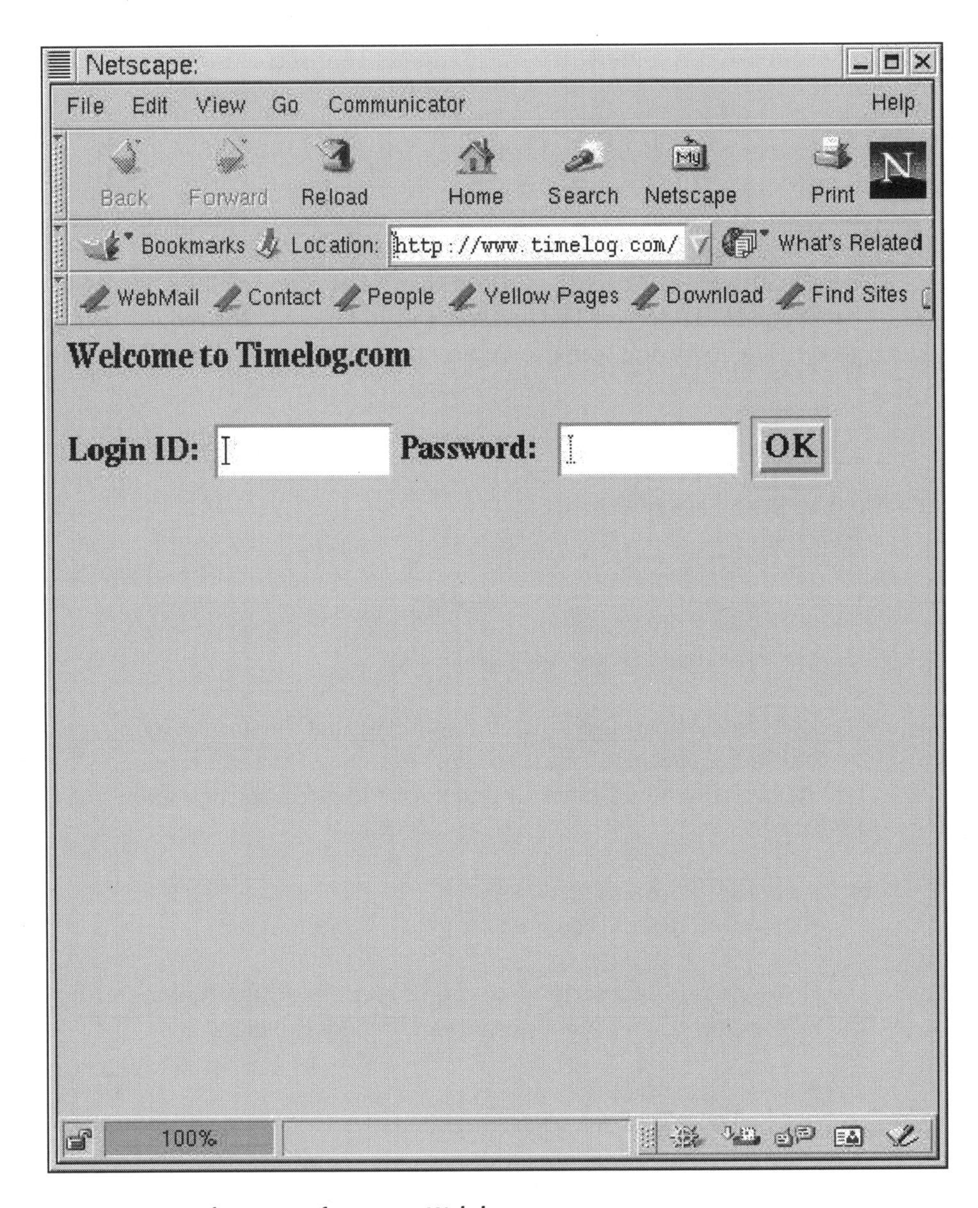

FIGURE 2.3 Login request form on a Web browser.

Repeat the Netscape command for all other domains, and view the home page.

```
# netscape http://www.project.com
# netscape http://www.digital.com
# netscape http://www.printer.com
```

Behind-the-Scenes HTTP Transactions

HTTP transactions begin with a client sending a connection request to the HTTP port (usually TCP/IP port 80) on the server. The browser then adds some text to the original request. If you enter "http://www.aklinux.com" as a request on the Web browser, such as Netscape, you are essentially establishing a connection to a remote server, nicknamed "www" in a fictitious domain "aklinux.com." Most likely, the server software is Apache, as more than 50 percent of ISPs use Apache as their backbone HTTP server. The browser will add more to the request being sent to the server. The following scenario depicts the behind-the-scenes HTTP transaction in a request and response sequence during a browser and server communication.

The Request

The request has three parts: "http:" is the protocol, "//www.aklinux.com" is the computer name in a domain "aklinux.com," and "index.html" is the requested document. The actual contents sent over the Internet are little more than that, and would look something like this:

```
1) GET /contributors/index.html HTTP/3.0
2) Connection: keep-Alive
3) User-Agent: Mozilla/4.5 [en] (X11; I; Linux 2.2.36 i586)
4) Host: www.aklinux.com
5) Accept: image/gif, image/x-xbitmap, image/jpeg, image/pjpeg,
   image/png, */*
6) Accept-Encoding: gzip
7) Accept-Language: en
8) Accept-Charset: iso-8859-3.*.utf-8
```

A blank line would follow this data, indicating end of request.

The important thing is the request type GET. The other most common request type is POST, but it is mainly used when form data are being sent back to the server, such as Web forms.

After the Request

The server responds with a request header that looks something like the browser. There could be several lines of information followed by a blank line indicating the end of the header. The contents requested (in our case, the file index.html contents) follow. A typical header looks something like this:

```
1) HTTP/3.1 200 OK
2) Date: Sun, 10 jun 2001 04:20:35 GMT
3) Server: Apache/3.3.5-dev (Unix) PHP/3.0.6
```

```
4)  Cache-Control: max-age=86400
5)  Expires: Mon, 11 jun 2001 04:20:35 GMT
6)  Connection: close
7)  Content-Type: text/html
```

Line 1 is the status code; 200 OK means requested documents were found. Lines 2 though 5 indicate the date and time, the server software type, and expiration information, respectively. Line 6 means that the server will close the connection after the contents have been transferred. Line 7 tells the browser what type of documents to expect. The browser will keep trying to read the data until the server closes the connection. The other option is *keep-alive*. In this case, the server specifies the content length in the header. Once the browser reads the required amount, it issues another request for remaining documents. The advantage of keep-alive is that the browser does not have to establish a new HTTP connection.

Forms in the HTML Documents

If the index.html document you received has the following contents, it means you have received a form:

```
1)  <FORM ACTION="http://www.project.com/cgi-bin/test.pl"
    method="post">
2)  Name: <INPUT NAME="name" TYPE=text SIZE=10>
3)  Password:<INPUT NAME="password" TYPE=password SIZE=10><BR>
4)  <INPUT TYPE=submit value="Enter">
5)  <INPUT TYPE=submit value="Cancel">
6)  </FORM>
```

It starts with FORM, followed by the Action to be taken when the Submit button is pressed; for example, what CGI script to execute (in this case, //www.project.com/cgi-bin/test.pl). "Post" is the method by which the variables are being returned. The form elements are next, followed by the terminating symbol </FORM>.

Sending and Receiving Cookies

If the document you received from the server has the following contents, it means you have received a cookie:

```
1)  HTTP/3.1 200 OK
2)  Date: Sun, 10 jun 2001 04:20:35 GMT
3)  Server: Apache/3.3.5-dev (Unix) PHP/3.0.6
4)  Cache-Control: max-age=86400
5)  Expires: Mon, 11 jun 2001 04:20:35 GMT
```

```
6) Connection: close
7) Set-Cookie: id=1234; path=/; domain=aklinux.com;
8) Content-Type: text/html
```

Cookies are sent to the browsers from the server as tiny objects. They are in the form of name=value pairs. They are returned to the server by the browser when requesting another URL from the same domain. Cookies provide identification information to the server.

SUMMARY

In this chapter, we implemented the DNS server, the Web server, and the SQL server on Linux, which completes our user interface requirement of the embedded system functional design specifications. These services provide the glue logic for the components we will be using in our embedded system design projects. Incorporating full-blown database and Web servers is not feasible when there are space constraints on the storage media, but we can use the concept as a guideline to modularize our design requirements. The concept of dividing the project development cycle into the user interface and the system interface using database as the coordinator is applicable in all situations. If your product is being used in an intranet environment, it is definitely wise to leverage Web technology. You have already seen how to put information on the Web by creating virtual hosting on the Linux server, which can save you both time and money in your project design cycle.

We covered the setup procedures for the following server modules:

- SQL
- DNS
- Web

3 Device Drivers

In this chapter

- Device Driver Essentials
- Device Driver Module
- Device Driver Construct

INTRODUCTION

In our scheme of embedded system design, we have a user interface, a system interface, and a database. The Internet is our user interface, and we saw an example of the Internet setup in the previous chapter. Now, we proceed with the first part of the system interface setup. Embedded systems almost always communicate with an external device, and a device driver is the fundamental source of communication with the device. In this chapter, we will develop a device driver for your embedded system design.

As you know, all computer programs have input, processing, and output paradigms. Application programs are mainly concerned with processing, while the details of input and output are left to the underlying operating system. Input and output hardware devices such as keyboards, disk drivers, and monitors can only be accessed through special programs called *device drivers*. Direct access to the devices is not permitted by the operating system, so an application program has to go through device drivers to gain access (Figure 3.1).

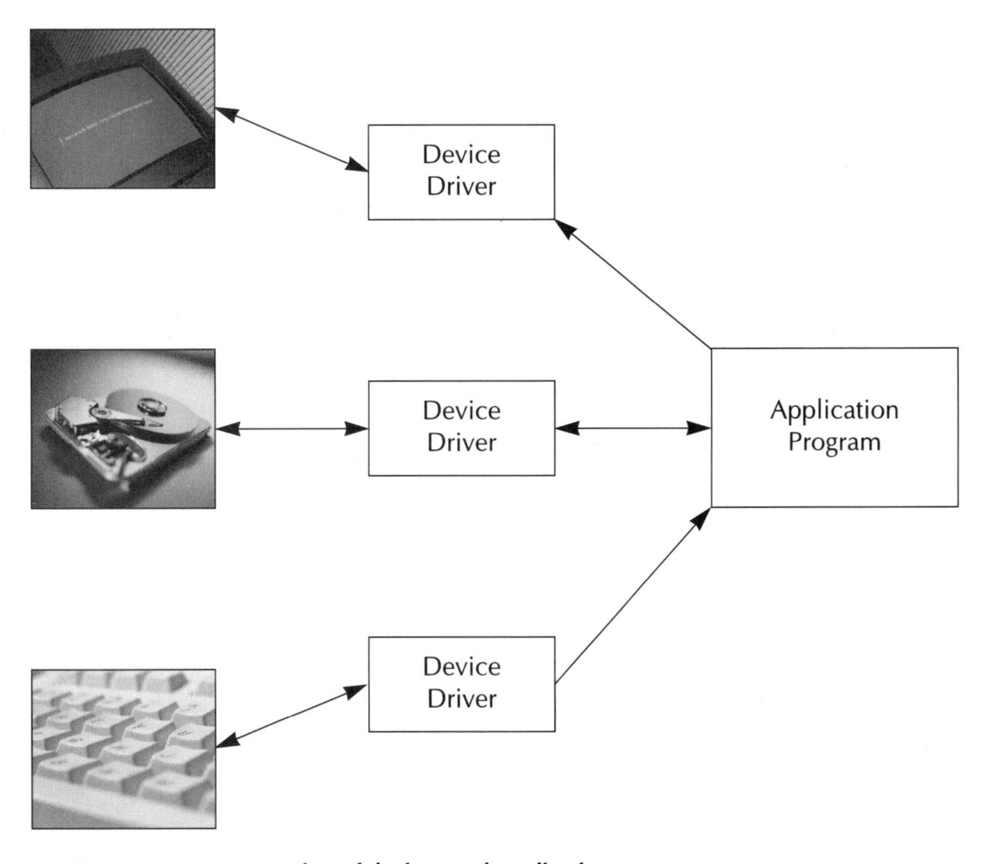

FIGURE 3.1 Interconnection of devices and applications.

Device driver programs are modules that are essentially part of the underlying operating system, and are executed inside the memory space of the operating system kernel. Linux treats all devices equally; it treats them as files. The most an application program can do is open, close, read, and write to these files. Some, such as keyboard and screen files, do not allow an open or close; they are always open.

There are two types of device drivers in the Linux kernel: *char* device drivers that operate one byte at a time, such as the keyboard and monitor; and *block* device drivers that operate multiple bytes at a time, such as a hard disk and CD-ROM.

Device driver modules have special permissions and privileges, as they are executed within the kernel's domain, and therefore have access to every bit of the kernel's code. There is a uniform format for all driver software that makes the coding style easier. As mentioned previously, for all practical purposes, the devices look like files to an application program. Almost all communication with the devices is performed via read and write operations, as if they were files on a hard disk. Just as the files on the hard disk are nodes in the file system, device drivers are also identified by unique numbers called *inodes*, also called *major* and *minor* numbers. These numbers are assigned statically by users or dynamically requested by the kernel, and should be unique.

The important difference between the device driver files and the other read and write files is that the device driver files appear in the /dev directory (with very few exceptions), and there is an entry in the /proc file system for each device driver in the system. /proc is a special file system for kernel usage only. The other differences are that the inodes for the device drivers must be known prior to their creation in the /dev directory. If we know the inode number beforehand (assigned statically by the user), then the order of creation—whether inserting into the /proc directory first or creating the entry in /dev directory first—is not important. However, if we ask the kernel to assign the inode number dynamically, we must insert the module into the /proc directory first and obtain the inode number from there that we can use to create an entry in the /dev directory.

We can summarize the creation of device drivers in Linux is two steps:

1. Insert the device driver module into the /proc file system.
2. Attach the file node that belongs to the device driver module with /dev directory.

DEVICE DRIVER ESSENTIALS

The first step in creating a driver is registering the driver with the operating system. Each driver is assigned a major and a minor number by the kernel that identifies its entry into the file system. The major number identifies the driver associated with the device, and the minor number is for internal use of the driver only. Adding a new driver to the system means assigning a major number to it at initialization time. The following function call assigns a major number to the char device driver:

```
#include <linux/fs.h>
int register_chrdev (unsigned int major, const char * name, struct
file_operations fops);
```

The major number ranges from 1 to 254 and is the index into a static array of the char drivers. You can force the kernel to use a predefined number, but you run the risk of running over some other modules that might have been assigned the same number and are already running in the system. In fact, the kernel reserves some major numbers for most common devices in the system.

It is better to ask the kernel to assign the major number dynamically, bypassing 0 for the parameter "major" in the register_chrdev() function.

Minimum Function Device Driver

ON THE CD

Whether you perform the file read and write operations or not, all device drivers must provide at least two functions: init_module and cleanup_module. The init_module is where we register the driver with the operating system, and the cleanup_module is where we "unregister" the driver from the kernel. The init_module is the first function called when the module is loaded into the kernel, and the cleanup_module is the last function executed in the life of the driver, before unloading it from the kernel. As a root user, create a file named pport.c in the directory /dd/minimum (the file is provided in the companion book CD directory /book/chapter3/minimum/pport.c), and insert the text shown in Listing 3.1. The program provides the following functionality:

- Lines 5 through 21 are the file_operation table; all the entries are blank, indicating that no file operation is supported.
- Lines 22 through 26 are the init_module function, registering the device with the kernel.
- Lines 27 through 30 are the cleanup_module function, unregistering the device from the kernel.

LISTING 3.1 File /book/chapter3/minimum/pport.c; a minimal device driver.

```
1)  #include <linux/module.h>
2)  #include <linux/kernel.h>
3)  #include <linux/fs.h>     /* everything... */
4)  int major = 0;
5)  struct file_operations fops = {
6)  NULL,          /* lseek */
7)  NULL,          /* read */
8)  NULL,          /* write */
9)  NULL,          /* readdir */
10) NULL,                          /* poll */
11) NULL,          /* ioctl */
12) NULL,          /* mmap */
13) NULL,          /* open */
14) NULL,                          /* flush */
15) NULL,          /* release */
16) NULL,          /* fsync */
17) NULL,          /* fasync */
```

```
18) NULL,                        /* check_media_change */
19) NULL,                        /* revalidate */
20) NULL,          /* lock */
21) };
22) int init_module(void)
23) {
24)    major = register_chrdev(0, "pport", &fops);
25)    return 0;
26) }
27) void cleanup_module(void)
28) {
29)    unregister_chrdev(major, "pport");
30) }
```

Device Driver Makefile

ON THE CD

Device drivers are linked with the kernel internal data structures and are loaded into the kernel memory space. There is no main() function in the device driver. The module must be compiled with the flags -D__KERNEL__ -DMODULE. As a root user, create a file named Makefile in the directory /dd/minimum (the file is provided in the companion book CD directory /book/chapter3/minimum/Makefile), and insert the text shown in Listing 3.2.

LISTING 3.2 Makefile to compile and link pport.c.

```
1)  INCLUDEDIR = /usr/include
2)  DEBFLAGS = -O2
3)  CFLAGS = -D__KERNEL__ -DMODULE -Wall $(DEBFLAGS)
4)  CFLAGS += -I$(INCLUDEDIR)
5)  OBJS = pport.o
6)  all: $(OBJS)
```

Next, compile the file pport.c using the Makefile with the following shell command:

```
[root@ns /dd/minimum]#  make
```

Verify that the file pport.o is being created.

```
[root@ns  /dd/minimum]#  ls -al pport.o
```

DEVICE DRIVER MODULE

As you can see, there is no *main()* in Listing 3.1. The device drivers are essentially loaded and executed into the kernel space rather than the user space (as opposed to normal application programs). The following */sbin/insmod* command will insert the pport.o module into the kernel space:

```
[root@ns  /dd/minimum]# /sbin/insmod –o pport pport.o
```

The parameters to the *insmod* command are the module name (pport) and the filename (pport).You can verify the presence of your driver by viewing the contents of the directory /proc/module.

```
[root@ns  /dd/minimum]# cat  /proc/modules | grep pport
```

You can also use the *lsmod* command to view a listing of the module.

```
[root@ns  /dd/minimum]# lsmod
```

You can see the major number assigned to the driver by catenating the /proc/devices file.

```
[root@ns  /dd/minimum]# cat /proc/devices | grep pport
254 pport
```

The following */sbin/rmmod* command will remove and unregister the pport.o module from the kernel space.

```
[root@ns  /dd/minimum]# /sbin/rmmod pport
```

Device Drivers and File Nodes

It is not enough to insert the device module with the *insmod* command. Although the kernel assigns a major and minor number (file nodes) when the modules are inserted, attaching the file nodes to the file system is a separate process. In this process, a name is associated with the file nodes, and through this name, the normal file-read and file-write operations are performed. Only then do the devices appear as files in the system. The devices are usually mounted in the /dev directory. A listing of the directory shows a "c" in the permission field for character devices, and a "b" for block devices.

```
[root@ns  /dd/minimum]# cd /dev
[root@ns /dev]# ls -al
```

A name must be inserted into the /dev directory and be associated with the major number in order for the devices to be recognized programmatically. Knowing the major and minor numbers beforehand makes it is easier to insert the device into the file system; otherwise, the numbers have to be extracted from the /proc/devices file system. As mentioned previously, from an application program's point of view, a device driver is simply a file. The following *mknod* command creates a file in the /dev directory, given the major and minor numbers of the device. For devices that have dynamically assigned file nodes, the major number must be obtained from the entries into the */proc/devices* file. Later, a script is developed that extracts the major number from the /proc/devices file. The following shell command creates a node in the /dev directory for the module pport.o:

```
[root@ns  /dd/minimum]# mknod /dev/pport c 254 0
```

The device driver pathname is /dev/pport; "c" is for the char driver, 254 is the major number, and 0 is the minor number. If device nodes are assigned dynamically, as suggested earlier during the registration process, the major number can be read from the list of /proc/modules.

A program can now open the device and perform read and write and close operations as follows:

```
.
.
FILE * fd = fopen ("/dev/pport", "r");
.
.
```

Before inserting a new node for the same driver, it is important to remove the old node numbers that might have been left behind. You can use the following command to remove a stale node:

```
[root@ns  /dd/minimum]# rm -f  /dev/pport
```

Script for Loading a Device Driver

ON THE CD

A script is a text file that is interpreted by the shell upon execution. Each line of text in the file is an individual command, as if it was entered through the keyboard. Thus, we can combine the steps we described in the previous section of loading a device driver into a script file as shown in Listing 3.3 (the file is provided in the companion book CD directory /book/chapter3/simple/pp_load). The script file automates the process of loading and unloading the modules. The following is an explanation of the steps in the script:

- Line 1 tells the command interpreter that the file should be interpreted by the shell.
- Lines 2 through 5 are initialized variables being used in the script.
- Lines 7 through 9 remove the previous module and insert a new one.
- Lines 10 and 11 extracts and prints out the lines with the device major number.
- Line 12 removes any existing device file with the same major number.
- Line 13 creates nodes for devices being loaded.
- Lines 14 and 15 change permissions on the device files.

LISTING 3.3 File /book/chapter3/minimum/pp_load; script to extract the node number from /proc and create /dev entry for the pport module.

```
1) #!/bin/sh
2) module="pport.o"
3) device="pport"
4) group="root"
5) mode="664"
6) #remove any existing node
7) /sbin/rmmod pport
8) # invoke insmod with all arguments we got
9) /sbin/insmod -f $module $device || exit 1
10) major='cat /proc/devices | awk "\\$2==\"$device\" {print \\$1}"'
11) echo "Major number = $major"
12) rm -f /dev/${device}
13) mknod /dev/${device} c $major 0
14) chgrp $group /dev/${device}
15) chmod $mode  /dev/${device}
```

Unloading the Driver from the System

It is important to remove the major number from the system once the driver is unloaded, as it will cause problems if the kernel accesses the driver function that is no longer in the system. The following command unloads the driver module pport from the system:

```
[root@ns /dd/minimum]# rmmod pport
```

The kernel calls the driver function cleanup_module() when a request is received to unload the module. It is cleanup_module() function's responsibility to make a call to get itself unregistered as follows:

```
int unregister_chrdev(unsigned int major, const char *name);
```

The major number is the one that was assigned at registration time, and the char *name is the name of the driver that is placed in the /proc/devices file.

Script for Unloading a Device Driver

ON THE CD

Similar to loading a device driver script file, the script file in Listing 3.4 unloads the pport modules (the file is provided in the companion book CD directory /book/chapter3/simple/pp_unload).

LISTING 3.4 File /book/chapter3/minimum/pp_unload; script to unload the pport device driver from the system.

```
1)  #!/bin/sh
2)  module="pport"
3)  device="pport"
4)  # invoke rmmod with all arguments we got
5)  /sbin/rmmod $module $* || exit 1
6)  # Remove stale nodes
7)  rm -f /dev/${device}
```

DEVICE DRIVER CONSTRUCT

The minimal device driver we presented earlier showed the mechanism for creating, loading, and unloading a device driver, but had no practical value. Although a file node was being created, we could not perform the regular file operations such as open, close, read, write, and so forth. In this section, we will add the capability of file operations to our device driver, and discuss the different structures that make up the file operations of the underlying operating system.

The "struct file" Structure

The kernel keeps all the important information concerning a device driver in one structure struct file. It is different from the structure FILE that is used by the user program when a file is opened. The important fields in the struct file are as follows:

mode_t f_mode; The file mode is identified by bits FMODE_READ and FMODE_WRITE.

loff_t f_pos; The current reading or writing position.

unsigned short f_flags; These are the FILE flags, such as O_RDONLY, O_NONBLOCK, etc.

struct inode * f_inode; The inode associated with open file. In case the kernel does not provide the inode structure explicitly, the real inode can be obtained here.

struct file_operations * f_op; See the next section.

void * private_data; The open system call sets this pointer to NULL before calling the open function in the module. The driver is free to use this pointer for any reason.

The "struct file_operations" Structure

This structure is essentially a jump table of functions that are invoked whenever an operation is performed on a file whose major number matches the driver major number. If a specific operation is not needed or not being implemented, you can simply provide NULL as the parameter. Sometimes, a function in the table is not applicable to the device for which the driver is being written, such as "lseek" and "check media change," which are applicable to the tape devices only. The example device driver only provides read, write, open, and close operations.

```
struct file_operations pp_fops = {
    NULL,
    NULL,           /* lseek */
    pp_read,
    pp_write,
    NULL,           /* readdir */
    NULL,           /* poll */
    NULL,           /* ioctl */
    NULL,           /* mmap */
    pp_open,
    NULL,           /* flush */
    pp_close,
    NULL,           /* pp_fsync */
    NULL,           /* pp_fasync */
    /* nothing more, fill with NULLs */
};
```

The following is a list of functions corresponding to the entry into the file_operations structure:

```
int (*read)  (struct indoe *, struct file *,char *, int);
int (*write)  (struct indoe *, struct file *,const char *, int);
int (*open)  (struct indoe *, struct file *,off_t, int);
int (*close)  (struct indoe *, struct file *,off_t, int);
int (*ioctl)  (struct indoe *, struct file *,unsigned int ,unsigned long);
```

The same functions in the user space are specified as follows:

```
FILE *fopen(const char *filename, const char *mode);
size_t fread (void *ptr, size_t size, size_t nobj, FILE *stream);
size_t fwrite (const void *ptr, size_t size, size_t nobj, FILE *stream);
int (*open)  (struct indoe *, struct file *, off_t, int);
int fclose  (FILE *fd);
int ioctl (int fd, int cmd, Ö);
```

The third argument in ioctl() is actually a single optional argument, and the dots are there to prevent type checking. In the pport driver, the third argument of the function ioctl() is a pointer that receives the actual status, as you will see in the test program that receives the parallel port status register.

Role of the Minor Number in a Device Driver

As mentioned earlier, the kernel is only concerned about the major number, but the combined data type "major and minor number" resides in the field i_rdev of the inode structure. Every driver function receives a pointer to the inode as the first argument. The function can extract the minor device number by looking at inode->i_rdev.

The following section describes the implementation of a device driver with a specific example using the parallel port of the PC platform. Once the device driver is in place, an application program can access all of the parallel port registers (data register, status register, and control register) by simply reading and writing as characters from a file. The read function returns the status register contents, the write function writes the byte value to the data register of the parallel port, and the ioctl function sets the control register of the parallel port.

A Full-Function Device Driver

ON THE CD

The struct file_operations identifies ten functions that the device driver exposes to the outside world; the most common are open, close, read, and write. The example parallel-port device driver (Listing 3.5) has been implemented with the common functions only. With this device driver, an application program will be able to monitor and control the logic level of the pin-outs of the parallel port connector (Figure 3.2). The write function writes to the output register (data register memory location 0x378), and the read function reads from the input register (status register memory location 0x379) of the PC parallel port. The ioctl function is to set and reset the ctrl register, which we will discuss later. The struct file_operations structure will be initialized with the address of the read, write, open, and close functions, and will be passed to the kernel during init_module function call at module load time. Listing 3.5 is a parallel-port device driver that provides access to the low-level registers of the parallel-port hardware (the file is provided in the companion book CD directory /book/chapter3/simple/pport.c). The following is an explanation of the program listing:

- Lines 6 through 14 are the include files required for definitions.
- Lines 15 through 17 are initialized variables. The value of pp_major is 0, requesting the kernel to provide the major number.
- Lines 19 through 25 are the open function.
- The macro in line 21, MOD_INC_USE_COUNT, indicates the device is currently in use.
- Line 22 simply saves the minor device number being passed by the kernel.
- printk in line 23 is a special printf function for kernel use only; the output appears in the /var/log/messages file.
- The function pp_close() in lines 27 through 31 is the close() function that simply decrements the use count; the driver should be removed only when the use count reaches 0.
- Lines 33 through 41 are the read() function, where line 37 gets the input value from the port (base 0x379), and line 38 copies it to the user memory area.
- Lines 43 through 51 is the write() function. Line 47 gets the input value from the parameter (let the kernel copy from the user memory area to the system memory area) and writes to the port (base 0x378).
- Lines 53 through 68 are the initialization of the file_operation structure.
- Lines 70 through 97 are the init_module function. Line 72 checks to see if the hardware addresses are available for use, and line 81 requests the kernel for the use of the hardware address range. Line 86 is the registration of the device driver, and line 95 requests a memory page from the kernel that cannot be swapped out to the disk.
- Lines 99 through 106 are the cleanup_module() function. The function will be called when the module is unloaded with the unregister function call.

LISTING 3.5 /book/chapter3/simple/pport.c device driver: open, close, read, and write functions.

```
1)     /**************************************************************/
2)     /*
3)     pport.c – Simple Hardware Operations Read and Write only
4)     *
5)     *********/
6)     #include <linux/module.h>
7)     #include <linux/sched.h>
```

```
8)    #include <linux/kernel.h> /* printk() */
9)    #include <linux/fs.h>     /* everything... */
10)   #include <linux/errno.h>  /* error codes */
11)   #include <asm/io.h>
12)   #include <linux/ioport.h>
13)   #include <asm/segment.h>
14)   #include <asm/uaccess.h>
15)   int pp_major = 0;
16)   int pp_base = 0x378; /* intel: default to lp0 */
17)   int minor = 0;
18)   /*********************************************************/
19)   int pp_open (struct inode *in, struct file *filp)
20)   {
21)       MOD_INC_USE_COUNT;
22)       minor =(MINOR(in->i_rdev)&0x0f);
23)       printk(KERN_INFO "pp:U OPEN %d\n", minor);
24)       return 0;
25)   }
26)   /*********************************************************/
27)   int pp_close (struct inode *in, struct file *filp)
28)   {
29)       MOD_DEC_USE_COUNT;
30)       return(0);
31)   }
32)   /*********************************************************/
33)   int pp_read (struct file * filp, char *buf, size_t count, loff_t * t)
34)   {
35)       unsigned char x;
36)       unsigned port = pp_base + minor;
37)       x = inb(port + 1);          /* Status reg offset */
38)       put_user(x,buf);
39)       printk(KERN_INFO "pp: READ %2.2x\n",x);
40)       return 1;
41)   }
42)   /*********************************************************/
43)   int pp_write (struct file * filp, const char *buf, size_t count, loff_t * t)
44)   {
45)       unsigned char x;
46)       unsigned port = pp_base + minor;
47)       get_user(x,buf);
48)       printk(KERN_INFO "pp: WRITE %2.2x\n",x);
49)       outb(x, port);
50)       return 1;
```

```
51)  }
52)  /******************* file_operations************************/
53)  struct file_operations pp_fops =
54)  {
55)      NULL,          /* lseek */
56)      pp_read,
57)      pp_write,
58)      NULL,          /* readdir */
59)      NULL,          /* poll */
60)      NULL,          /* ioctl */
61)      NULL,          /* mmap */
62)      pp_open,
63)      NULL,          /* flush */
64)      pp_close,
65)      NULL,          /* pp_fsync */
66)      NULL,          /* pp_fasync */
67)     /* nothing more, fill with NULLs */
68)  };
69)  /********************* init_module *************************/
70)  int init_module(void)
71)  {
72)      int result = check_region(pp_base,4);
73)      #ifdef DEBUG
74)      printk(KERN_INFO "pp: INIT_MOD\n");
75)      #endif
76)      if (result)
77)      {
78)           printk(KERN_INFO "pp: can't get I/O address 0x%x\n",pp_base);
79)           return result;
80)      }
81)      if (!request_region(pp_base,4,"pp"))
82)      {
83)           printk(KERN_INFO "pp: port %x is busy\n",pp_base);
84)           return 0;
85)      }
86)      result = register_chrdev(pp_major, "pp", &pp_fops);
87)      if (result < 0)
88)      {
89)          printk(KERN_INFO "pp: can't get major number\n");
90)          release_region(pp_base,4);
91)          return result;
92)      }
93)      if (pp_major == 0)
```

```
94)            pp_major = result; /* dynamic */
95)        pp_buffer = __get_free_page(GFP_KERNEL); /* never fails */
96)        pp_head = pp_tail = pp_buffer;
97)    }
98)    /*********************** cleanup_module ******************/
99)    void cleanup_module(void)
100)   {
101)       unregister_chrdev(pp_major, "pp");
102)       release_region(pp_base,4);
103)       if (pp_buffer)
104)           free_page(pp_buffer
105)   }
106)   /*********************************************************/
```

Testing the Device Driver

A test scenario is provided in the following section that would verify the functionality of the pport device driver's read and write methods.

Test Objective

The objective of this test is to test the pport device driver read and write operations.

Test Description

This will test the pport device driver's ability to set and reset the pins on the external parallel port connector by controlling the contents of the parallel port data register (address 0x378), and read the status of the input pins by reading the status register (address 0x379) (Figure 3.2). If successful, the command line will display on the screen the contents of the status register upon execution of the test program.

Test Preparation

The schematic in Figure 3.2 shows the pin-outs of the parallel port connector. The status port is a read-only register with the data bit -7 as an inverted bit. If any of the output pins of the data register (DB25 pins 2, 3, 4, 5, 6, 7, 8, 9) are connected with any of the input pins of the status register (DB25 pins 15, 13, 10, 11), the contents you write will appear on the status register as loop back. In the first step, connect pin 2 (data bit -0) to pin 11 (status bit 7).

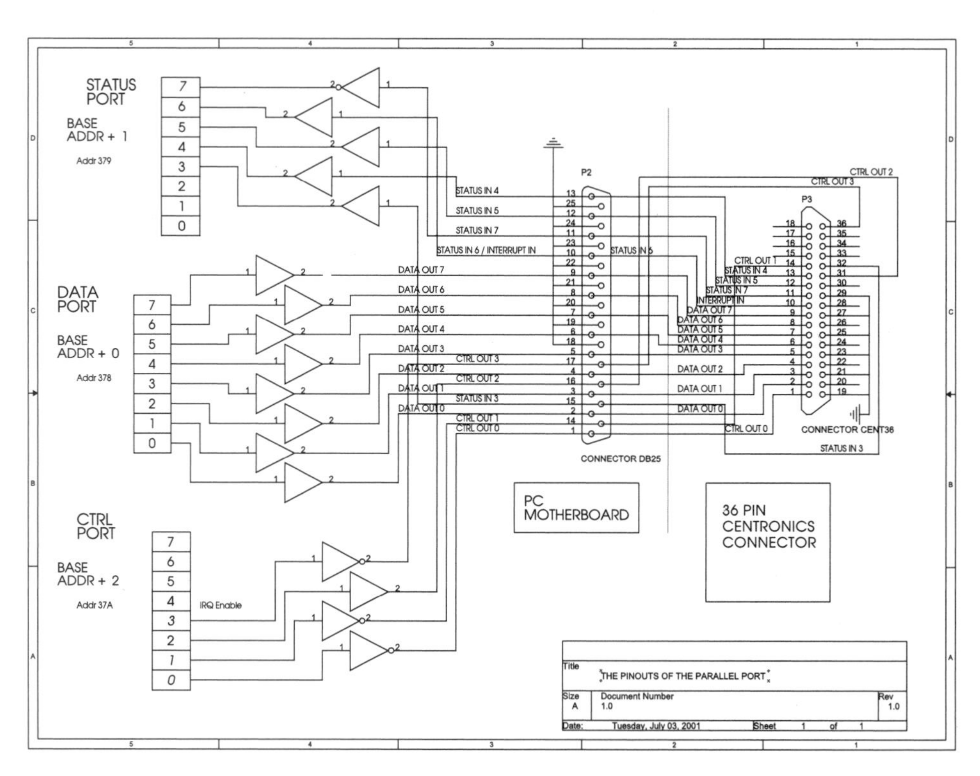

FIGURE 3.2 Schematic of a PC platform parallel port with Centronics connector.

Test Input

1. Mount the CD-ROM if you are planning to copy the source code from the CD.

```
mount -t iso9660 /de/cdrom /mnt/cdrom
```

ON THE CD

2. Create a subdirectory /dd/simple, and edit the device driver source file /dd/simple/pport.c as defined in Listing 3.5, or copy the source file from the CD-ROM directory /book/chapter3/simple/pport.c.

```
mkdir /dd/simple
cp /mnt/cdrom/book/chapter3/simple/pport.c  /dd/simple/.
```

ON THE CD

3. Edit the device driver Makefile /dd/simple/Makefile as defined in Listing 3.2, or copy the source file from the CD-ROM directory /book/chapter3/simple/Makefile.

```
[root@ns /dd/simple]# cp /mnt/cdrom//book/chapter3/simple/Makefile  /dd/simple/.
```

4. Compile and link the pport.c using the make utility.

```
[root@ns /dd/simple]# /dd/simple/make
```

ON THE CD

5. Edit the script to load the device driver as defined in Listing 3.3, or copy it from the CD-ROM directory /book/chapter3/simple/simple/pp_load.

```
[root@ns /dd/simple]# cp /mnt/cdrom/dd/simple/pp_load  /dd/simple/.
```

6. Load the device driver using the pp_load script.

```
[root@ns /dd/simple]# /dd/simple/pp_load
```

7. Create a subdirectory, /dd/simple/test, and edit the file /dd/simple/test/test.c as shown in Listing 3.6, or copy it from the CD-ROM directory /dd/simple/test/* using the following command:

```
[root@ns /dd/simple]# cp /mnt/cdrom/dd/simple/test/*  /dd/simple/test/.
```

An explanation of the test program follows:

- Lines 5 through 22 are the loopback test function; the input to the function is the bit pattern to be written to the parallel port data register.
- Lines 24 through 50 are the main function. After printing a banner, the program gets into the forever loop (lines 29 through 49) requesting the input pattern and calling the loopback test function.

LISTING 3.6 File test.c to test the pport device driver.

```
1)    #include <stdio.h>
2)    #include <stdlib.h>
3)    #include <fcntl.h>
4)    #include <unistd.h>
5)    /*****************************/
6)    void LoopBackTest(int pattern)
7)    {
```

```
8)      int fh, cnt;
9)      char bfr[2];
10)     bfr[0] = (char) pattern;
11)     printf("LoopBack test begin %2.2x\n",bfr[0]);
12)     fh = open("/dev/pport",O_RDWR);
13)     if (fh)
14)     {
15)        write(fh,bfr,1);
16)        cnt = read(fh,bfr,1);
17)        printf("Status Port bits %2.2x\n",bfr[0] & 0xff);
18)        close(fh);
19)     }
20)     else
21)        printf("Fail open device\n");
22)  }
23)  /*****************************/
24)  int main(int argc , char *argv[])
25)  {
26)     int pattern;
27)     printf("Loop Back test for the parallel port device driver \n");
28)     printf("Press Ctrl C to exit\n");
29)     while(1)
30)     {
31)        printf ("Data register\n");
32)        printf ("bit 0 ——> DB25 pin 2\n");
33)        printf ("bit 1 ——> DB25 pin 3\n");
34)        printf ("bit 2 ——> DB25 pin 4\n");
35)        printf ("bit 3 ——> DB25 pin 5\n");
36)        printf ("bit 4 ——> DB25 pin 6\n");
37)        printf ("bit 5 ——> DB25 pin 7\n");
38)        printf ("bit 6 ——> DB25 pin 8\n");
39)        printf ("bit 7 ——> DB25 pin 9\n");
40)        printf ("Status register (Notice bit 0,1,2 are not
           available)\n");
41)        printf ("bit 3 <—— DB25 pin 15\n");
42)        printf ("bit 4 <—— DB25 pin 13\n");
43)        printf ("bit 5 <—— DB25 pin 12\n");
44)        printf ("bit 6 <–Inverted— DB25 pin 10\n");
45)        printf ("bit 7 <—— DB25 pin 11\n");
46)        printf("Enter bit pattern in hex for data register ");
47)        scanf ("%x",&pattern);
48)        LoopBackTest(pattern);
49)     }
50)  }
```

ON THE CD Edit the Makefile file /dd/simple/test/Makefile as shown in Listing 3.7, or copy it from the CD-ROM directory /book/chapter3/simple/test/Makefile. Compile and link the test.

```
[root@ns /dd/simple/test]#  make
```

LISTING 3.7 /book/chapter3/simple/test/Makefile for the test program.

```
1)  all: t
2)  CC = gcc
3)  INCLUDEDIR = /usr/include
4)  CFLAGS = -g -02 -Wall -I$(INCLUDEDIR)
5)  t: test.o
6)  $(CC) -o t test.o
7)  test.o: test.c
8)  $(CC) -c test.c $(CFLAGS)
```

8. Change the mode of the output file t to make it executable.

```
[root@ns /dd/simple/test]# chmod 644 /dd/simple/test/t
```

Expected Test Results

The program will request a byte pattern to be written to the data register. In return, a byte pattern received from the status register will be displayed. If the loopback connector is correct, you should see the inverted value of the bit displayed on the screen.

Test Procedure

1. From the command line, execute the program t.

```
# ./t
```

2. Enter the byte pattern at the command prompt.

```
Please enter a new byte pattern 0x55
0x80
Please enter a new byte pattern 0xAA
0x00
```

Test Cleanup

ON THE CD

Press Ctrl-C to exit the program, and unload the device driver with the pp_unload script as defined in Listing 3.4. The script is also available from the CD-ROM directory. /dd/simple/pp_unload

```
[root@ns /dd/simple]#  /dd/simple/unpp_load
```

Adding ioctl Functionality

The fifth entry in the file_operations is the ioctl function call. This is the routine for everything that you cannot accomplish with the normal read and write functions, such as setting up hardware registers, changing the mode of the device, and so forth. The ioctl implementation is strictly device dependent. To continue with our parallel port device driver, we will add the functionality of reading and writing of the control register through ioctl function calls. Listing 3.8 is the section of the code that will be added to the source code pport.c in Listing 3.5. The following is an explanation of the additions to the source list:

- Add the ioctl numbers to our pport device driver (lines 1 through 2) to gain access to the parallel port ctrl register (address 0x37A).
- Line 30 is an entry for ioctl function to the struct file_operations.
- Lines 3 through 21 are a function declaration to handle the ioctl call.

LISTING 3.8 The code section added to the pport driver to handle ioctl calls.

```
1)  #define GET_CTRL_REG                    1
2)  #define SET_CTRL_REG                    2
3)  int pp_ioctl (struct inode *inode, struct file *filp, unsigned int
    cmd, unsigned long arg)
4)  {
5)     unsigned char x;
6)     switch(cmd)
7)     {
8)     case GET_CTRL_REG:
9)         x = inb(pp_base + 2); /* ctrl reg offset */
10)        put_user(x,buf);
11)        printk(KERN_INFO "pp: IOCTL GET_CTRL_REG %2.2x\n",x);
12)        break;
13)    case SET_CTRL_REG:
14)        get_user((pp_base + 2),x); /* ctrl reg offset */
15)        printk(KERN_INFO "pp: IOCTL SET_CTRL_REG %2.2x\n",x);
16)        outb(x, port);
```

```
17)         break;
18)     default:
19)     }
20)     return 0;
21) }
22) /************************
    file_operations*****************************/
23) struct file_operations pp_fops =
24) {
25)     NULL,           /* lseek */
26)     pp_read,
27)     pp_write,
28)     NULL,           /* readdir */
29)     NULL,           /* poll */
30)     pp_ioctl,       /* ioctl */
31)     NULL,           /* mmap */
32)     pp_open,
33)     NULL,           /* flush */
34)     pp_close,
35)     NULL,           /* pp_fsync */
36)     NULL,           /* pp_fasync */
37)     /* nothing more, fill with NULLs */
38) };
```

Testing the ioctl Functionality

In this test scenario, you would verify the ioctl functionality of the pport device driver.

Test Objective

The objective of this test is to test the pport device driver's ioctl function implementation.

Test Description

This will test the pport device driver's ability to set and reset the control pins on the external parallel port connector (DB25 pins 1, 14, 16, 17). The ctrl register (address 0x37A) bit -4 is being used to enable or disable the interrupt input of the parallel port (DB25 pin 10). A loopback connection of the four ctrl register output pins with the status register input pin will enable us to read back the contents of the ctrl register. Figure 3.2 shows the ctrl register bits 0, 1, 3 as inverted, and bit 2 is not inverted. If successful, executing the test program will display the contents of the status register reflecting the ctrl register bit setting.

Test Preparation

Figure 3.2 shows the pin-outs of the parallel port connector. Connect pin 1 (ctrl register bit 0) with the DB25 pin 11 (status register bit 7). A write to the ctrl register will loop back to the read of the status register.

Test Inputs

1. Mount the CD-ROM if you are planning to copy the source code from the CD.

```
# mount -t iso9660 /dd/cdrom /mnt/cdrom
```

2. Create a subdirectory /dd/ioctl, and edit the device driver source file /dd/ioctl/pport.c as defined in Listing 3.8, or copy the source file from the CD-ROM directory /book/chapter3/ioctl/pport.c.

```
# mkdir /dd/ioctl
# cp /mnt/cdrom/book/ioctl/pport.c  /dd/ioctl/
```

3. Edit the device driver Makefile /dd/ioctl/Makefile as defined in Listing 3.2, or copy the source file from the CD-ROM directory /book/chapter3/ioctl/Makefile.

```
# cp /mnt/cdrom/book/chapter3/ioctl/Makefile  /dd/ioctl/
```

4. Compile and link the pport.c using the make utility.

```
# /dd/ioctl/make
```

5. Edit the script to load the device driver as defined in Listing 3.3, or copy it from the CD-ROM directory /book/chapter3/ioctl/pp_load.

```
# cp /mnt/cdrom/book/chapter3/ioctl/pp_load  /dd/ioctl/
```

6. Load the device driver using the pp_load script.

```
#  /dd/ioctl/pp_load
```

The test program for testing the ioctl functionality is similar to the simple device driver, except for the ioctl call in line 9 of Listing 3.9. Create a subdirectory /dd/ioctl/test, and edit the file /dd/ioctl/test/test.c as shown in Listing 3.9, or copy it from the CD-ROM directory /book/chapter3/ioctl/test/test.c. The following is an explanation of the test program:

- Lines 9 through 26 are the loopback test function; the input to the function is the bit pattern to be written to the parallel port ctrl register.
- Lines 28 through 50 are the main function. After printing a banner, the program gets into the forever loop (lines 33 through 49) requesting input pattern and calling the loopback test function.
- Repeat the test for bits 1, 2, and 3 of the ctrl register.

LISTING 3.9 File /book/chapter3/Ioctl/test/test.c to test the pport device driver ioctl facility.

```
1   #include <stdio.h>
2)  #include <stdlib.h>
3)  #include <fcntl.h>
4)  #include <unistd.h>
5)  #include <sys/ioctl.h>
6)  #define CTRL_PORT_WR 1001
7)  #define CTRL_PORT_RD 1002
8)  /*****************************/
9)  void LoopBackTest(int pattern)
10) {
11)    int fh;
12)    char bfr[2];
13)    printf("LoopBack test begin\n");
14)    fh = open("/dev/pport",O_RDWR);
15)    if (fh)
16)    {
17)       bfr[0] = pattern;
18)       printf("Pattern %2.2x\n",bfr[0]&0xff);
19)       ioctl (fh, CTRL_PORT_WR,bfr);
20)       ioctl (fh, CTRL_PORT_RD,bfr);
21)       printf("Ctrl Port bits %2.2x\n",bfr[0]&0xff);
22)       close(fh);
23)    }
24)    else
25)       printf("Fail open device\n");
26) }
27) /*****************************/
28) int main(int argc , char *argv[])
29) {
30)    int pattern;
```

```
31)   printf("Loop Back test for the parallel port device driver \n");
32)   printf("Press Ctrl C to exit\n");
33)   while(1)
34)   {
35)      printf ("Status register\n");
36)      printf ("bit 3 <—— DB25 pin 15\n");
37)      printf ("bit 4 <—— DB25 pin 13\n");
38)      printf ("bit 5 <—— DB25 pin 12\n");
39)      printf ("bit 6 <—Inverted— DB25 pin 10\n");
40)      printf ("bit 7 <—— DB25 pin 11\n");
41)      printf("Control register\n");
42)      printf("bit 0 <—-Inverted—-> DB25 pin 1\n");
43)      printf("bit 1 <—-Inverted—-> DB25 pin 14\n");
44)      printf("bit 2 <————————> DB25 pin 16\n");
45)      printf("bit 3 <—-Inverted—-> DB25 pin 17\n");
46)      printf("Enter the bit pattern in hex for loop back test ");
47)      scanf ("%x",&pattern);
48)      LoopBackTest(pattern);
49)   }
50) }
```

ON THE CD

7. Edit the Makefile file /dd/ioctl/test/Makefile as shown in Listing 3.7, or copy it from the CD-ROM directory /book/chapter3/ioctl/test/Makefile. Compile and link the test.

```
#  make
```

8. Change the mode of the output file t to make it executable.

```
# chmod 644 /dd/ioctl/test/t
```

Expected Test Results

The program will request a byte pattern to be written to the ctrl register. In return, a byte pattern received from the status register will be displayed; if the loopback is correct, you should see the bit setting displayed on the screen.

Test Procedure

1. From the command line, execute the program t.

```
# ./t
```

2. Enter the byte pattern at the command prompt.

```
Please enter a new byte pattern 0x55
0x55
Please enter a new byte pattern 0xAA
0xAA
```

Test Cleanup

ON THE CD

Press Ctrl-C to exit the program, and unload the device driver with pp_unload script as defined in Listing 3.4. The script is also available from the CD-ROM directory /book/chapter3/ioctl/pp_unload

```
#  /dd/ioctl/pp_unload
```

Adding Interrupt Handling

ON THE CD

The PC parallel port can be programmed to generate an IRQ request. A low-to-high transition on the DB25 connector pin 10 generates an interrupt request that is vectored at IRQ 7. To handle the interrupt request, we can improve our device driver to recognize the event. Listing 3.10 is a modified version of Listing 3.5 that includes interrupt handling. The interrupts are asynchronous events that are happening in real time. The PC platform uses the parallel port interrupt to receive acknowledgment from the printer device that a character has been received and it is now ready to receive the next character. In our example software, we will use the parallel port interrupt to measure the pulse width, or a tachometer that measures the RPM of a motor, or simply an event counter. The following is an explanation of the additions to Listing 3.10 for interrupt handling capability to the device driver software. The file is available in the companion CD directory path /book/chapter3/interrupt/pport.c.

- Lines 18 through 20 add three more ioctl numbers: WAIT_ON_INTERRUPT, INTERRUPT_PULSE_WIDTH, and INTERRUPT_EVENT_COUNTER.
- Lines 27 through 30 add variables to handle irq events. The pp_buffer points to the allocated kernel memory page where the timestamps of the IRQ events are being stored in a round-robin buffer. The pp_head points to the next available slot of the buffer, and pp_tail is being used as the pointer that is being read by the application program.
- Lines 127 through 148 are the interrupt handler function pp_interrupt(). This function is being called when a low-to-high transition appears on the IRQ line (pin 10 of parallel port DB25 connector). The function gets the current time of day (line 132) and saves it to the buffer pointer pp_head. The pointer is incre-

mented, and if it exceeds the limits (PAGE_SIZE), it wraps around and starts from the beginning (lines 143 through 144). The event count is incremented by 1 (line 145).

- Lines 74 through 83 are the ioctl handler for the INTERRUPT_PULSE_WIDTH. The pp_tail points to the next timeval structure that would be returned upon request.
- Lines 84 through 88 are the ioctl handler for the INTERRUPT_EVENT_COUNTER that returns the value of the event_counter variable.
- Lines 89 through 96 are the ioctl handler for the WAIT_ON_INTERRUPT that returns only when an interrupt has occured.
- Line 169 installs pp_interrupt as the function to handle parallel port irq.
- Line 187 frees up the irq line.

LISTING 3.10 /book/chapter3/Interrupt/pport.c device driver with open, close, read, write, ioctl, and interrupts.

```
/*********
* pport.c - The Parallel Port Device Driver
*********/
1)    #include <linux/module.h>
2)    #include <linux/sched.h>
3)    #include <linux/wait.h>
4)    #include <linux/kernel.h> /* printk() */
5)    #include <linux/fs.h>     /* everything... */
6)    #include <linux/errno.h>  /* error codes */
7)    #include <linux/malloc.h>
8)    #include <linux/mm.h>
9)    #include <linux/ioport.h>
10)   #include <linux/interrupt.h>
11)   #include <linux/tqueue.h>
12)   #include <linux/time.h>
13)   #include <asm/io.h>
14)   #include <asm/segment.h>
15)   #include <linux/poll.h>
16)   #define CTRL_PORT_WR 1001
17)   #define CTRL_PORT_RD 1002
18)   #define WAIT_ON_INTERRUPT 1003
19)   #define INTERRUPT_PULSE_WIDTH 1004
20)   #define INTERRUPT_EVENT_COUNTER 1005
21)   static int pp_major = 0;
22)   static int pp_base = 0x378; /* intel: default to lp0 */
23)   static int pp_irq = 7;
24)   static unsigned long pp_buffer = 0;
```

```
25)  static int minor = 0;
26)  static int irqCount = 0;
27)  static unsigned long volatile pp_head;
28)  static volatile unsigned long pp_tail;
29)  static wait_queue_head_t pp_queue;
30)  static struct timeval current_time;
31)  /* struct wait_queue *pp_queue = NULL; */
32)  int pp_open (struct inode *inode, struct file *filp)
33)  {
34)     MOD_INC_USE_COUNT;
35)     minor =(MINOR(inode->i_rdev)&0x0f);
36)     printk(KERN_INFO "pp:U OPEN %d\n", minor);
37)     return 0;
38)  }
39)  int pp_close (struct file *filp)
40)  {
41)     MOD_DEC_USE_COUNT;
42)     return(0);
43)  }
44)  int pp_read (struct file * filp, char *buf, size_t count, loff_t * t)
45)  {
46)     unsigned char *kbuf=kmalloc(16, GFP_KERNEL);
47)     kbuf[1]='\0';
48)     kbuf[0] = inb(pp_base+1);
49)     /*    printk(KERN_INFO " READ VALUE pp: %x\n",*kbuf); */
50)     copy_to_user(buf, kbuf, 1);
51)     kfree(kbuf);
52)     return (0);
53)  }
54)  int pp_write (struct file * filp, const char *buf, size_t count,
     loff_t *t)
55)  {
56)     int retval = count;
57)     unsigned char *kbuf=kmalloc(16, GFP_KERNEL);
58)     if (!kbuf)
59)     return -ENOMEM;
60)     copy_from_user(kbuf, buf, 1);
61)     /* printk(KERN_INFO "pp: WRITE  %x, %x\n",kbuf[0], kbuf[1]);  */
62)     outb(kbuf[0], pp_base);
63)     kfree(kbuf);
64)     return retval;
65)  }
66)  int pp_ioctl (struct inode *Inode, struct file *filp, unsigned int
     cmd, unsigned long arg)
```

```
67)  {
68)     struct timeval tv;
69)     unsigned char *kbuf=kmalloc(16, GFP_KERNEL);
70)     if (!kbuf)
71)     return -ENOMEM;
72)     switch(cmd)
73)     {
74)     case INTERRUPT_PULSE_WIDTH:
75)        /* interrupt pulse width as calculated by irq service routinr */
76)        if (pp_head != pp_tail)
77)        {
78)           copy_to_user((char *) arg, (char *) pp_tail, sizeof
              (unsigned int));
79)           pp_tail = pp_tail + sizeof(unsigned int);
80)           if (pp_tail >= pp_buffer + PAGE_SIZE)
81)           pp_tail = pp_buffer; /* wrap around */
82)        }
83)        break;
84)     case INTERRUPT_EVENT_COUNTER:
85)        /* interrupt counts are being copied to arg pointer */
86)        copy_to_user((char *) arg, (char *) &irqCount, sizeof (long));
87)        irqCount = 0;
88)        break;
89)     case WAIT_ON_INTERRUPT:
90)        /* wait until an event occurs */
91)        init_waitqueue_head  (&pp_queue);
92)        interruptible_sleep_on(&pp_queue);
93)         /* fill in the time interrupt occurred */
94)        do_gettimeofday(&tv);
95)        copy_to_user((char *) arg, (char *) &tv, sizeof (struct
           timeval));
96)        break;
97)     case CTRL_PORT_WR:
98)        copy_from_user(kbuf, (char *) arg, 1);
99)        printk(KERN_INFO "pp: CTRL_PORT_WR  %x\n",kbuf[0]&0xff);
100)       outb(kbuf[0],pp_base+2);
101)       break;
102) case CTRL_PORT_RD:
103)    kbuf[0] = inb(pp_base+2);
104)    printk(KERN_INFO "pp: CTRL_PORT_RD  %x\n",kbuf[0]&0xff);
105)    copy_to_user((char *) arg, kbuf, 1);
106)    break;
107)    default:
108)    }
```

```
109)    kfree(kbuf);
110)    return 0;
111) }
112) struct file_operations pp_fops = {
113) NULL,
114) NULL,           /* pp_lseek */
115) pp_read,
116) pp_write,
117) NULL,           /* pp_readdir */
118) NULL,           /* poll */
119) pp_ioctl,       /* ioctl */
120) NULL,           /* mmap */
121) pp_open,
122) pp_close,
123) NULL,           /* pp_fsync */
124) NULL,           /* pp_fasync */
125) /* nothing more, fill with NULLs */
126) };
127) void pp_interrupt(int irq, void *dev_id, struct pt_regs *regs)
128) {
129)    struct timeval tv;
130)    int usec, sec;
131)    unsigned * timediff;
132)    do_gettimeofday(&tv);
133)    /* calculate micro second difference */
134)    usec = tv.tv_usec - current_time.tv_usec;
135)    sec = tv.tv_sec - current_time.tv_sec;
136)    if (usec < 0)
137)        usec += 1000000; /* adjusted for rollover */
138)    current_time.tv_usec = tv.tv_usec;
139)    current_time.tv_sec  = tv.tv_sec;
140)    timediff = (unsigned *) pp_head;
141)    timediff = (sec * 1000000) + usec; /* save pulse width in usec */
142)    pp_head = pp_head + sizeof(unsigned int);
143)    if (pp_head == pp_buffer + PAGE_SIZE)
144)    pp_head = pp_buffer; /* wrap */
145)    irqCount++;
146)    printk(KERN_INFO “pp: INTERPT  %d\n”,irqCount);
147)    wake_up_interruptible(&pp_queue);    /* awake any reading process
       */
148) }
149) int init_module(void)
150) {
151)    int result = check_region(pp_base,4);
```

```
152)     printk(KERN_INFO "pp: INIT_MOD\n");
153)     if (result)
154)     {
155)        printk(KERN_INFO "pp: can't get I/O address 0x%x\n",pp_base);
156)        return result;
157)     }
158)     request_region(pp_base,4,"pport");
159)     result = register_chrdev(pp_major, "pport", &pp_fops);
160)     if (result < 0)
161)     {
162)         printk(KERN_INFO "pp: can't get major number\n");
163)         release_region(pp_base,4);
164)         return result;
165)     }
166)     if (pp_major == 0) pp_major = result; /* dynamic */
167)     pp_buffer = __get_free_page(GFP_KERNEL); /* never fails */
168)     pp_head = pp_tail = pp_buffer;
169)     result = request_irq(pp_irq, pp_interrupt,
170)     SA_INTERRUPT, "pport", NULL);
171)     if (result)
172)     {
173)        printk(KERN_INFO "pp: can't get assigned irq %i\n", pp_irq);
174)        pp_irq = -1;
175)     }
176)     else
177)     { /* actually enable it – assume this *is* a parallel port */
178)        printk(KERN_INFO "INTERRUPT ON %i\n", pp_irq);
179)        outb(0x10,pp_base+2);
180)     }
181)     init_waitqueue_head  (&pp_queue);
182)     do_gettimeofday(&current_time);
183)     return 0;
184) }
185) void cleanup_module(void)
186) {
187)     free_irq(pp_irq, NULL);
188)     unregister_chrdev(pp_major, "pport");
189)     release_region(pp_base,4);
190)     if (pp_buffer)
191)        free_page(pp_buffer);
192) }
```

Testing the Interrupt Functionality

In this test scenario the interrupt handling of the pport device driver is being tested.

Test Objective

The objective of this test is to test the pport device driver's interrupt handling capability through WAIT_ON_INTERRUT, INTERRUPT_PULSE_WIDTH, and INTERRUPT_EVENT_COUNTER ioctl calls.

Test Description

This will test the pport device driver's ability to recognize interrupts on the DB25 pin 10 of the parallel port connector. If successful, upon execution of the test program, the command line will sequentially display the interrupt timestamps during a five-second period. It will also display a count of interrupt events that occurred during the same five-second period.

Test Preparation

Figure 3.2 shows the pin-outs of the parallel port connector. If the output pin 9 (data bit 7) is connected to pin 10 (status in 6/interrupt), an interrupt will be generated and the kernel will dispatch the function pp_interrupt(). To perform the test, the data bit 7 (DB25 pin 9) of the data register should be set high beforehand.

Test Input

1. Mount the CD-ROM if you are planning to copy the source code from the CD.

```
# mount –t iso9660 /de/cdrom /mnt/cdrom
```

2. Create a subdirectory /dd/intrpt, and edit the device driver source file /dd/intrpt/pport.c as defined in Listing 3.8, or copy the source file from the CD-ROM directory /book/chapter3/interrupt/pport.c.

```
# mkdir /dd/interrupt
# cp /mnt/cdrom/book/chapter3/interrupt/pport.c  /dd/interrupt/.
```

3. Edit the device driver makefile /dd/intrpt/Makefile as defined in Listing 3.2, or copy the source file from the CD-ROM directory /book/chapter3/interrupt/Makefile.

```
# cp /mnt/cdrom/book/chapter3/interrupt/Makefile  /dd/interrupt/
```

4. Compile and link the pport.c using the make utility.

```
# /dd/interrupt/make
```

5. Edit the script to load the device driver as defined in Listing 3.3, or copy it from the CD-ROM directory /book/chapter3/interrupt/pp_load.

```
# cp /mnt/cdrom/book/chapter3/interrupt/pp_load  /dd/interrupt/
```

6. Load the device driver using the pp_load script.

```
#  /dd/interrupt/pp_load
```

7. Create a subdirectory /dd/intrpt/test, and edit the file /dd/intrpt/test/test.c as shown in Listing 3.11, or copy it from the CD-ROM directory /book/chapter3/interrupt/test/test.c. The test program is a forever loop structure. It reads from the command line one byte at a time and writes to the data register of the parallel port, and then reads from the status line and dumps the contents on the screen.

```
# cp /mnt/cdrom/book/chapter3/interrupt/test/*  /dd/intrpt/test/.
```

LISTING 3.11 File /book/chapter3/Interrupt/test/test.c. Test the pport device driver interrupt facility.

```
1    #include <stdio.h>
2)   #include <stdlib.h>
3)   #include <fcntl.h>
4)   #include <unistd.h>
5)   #include <string.h>
6)   #include <sys/ioctl.h>
7)   #define CTRL_PORT_WR 1001
8)   #define CTRL_PORT_RD 1002
9)   #define WAIT_ON_INTERRUPT 1003
10)  #define INTERRUPT_PULSE_WIDTH 1004
11)  #define INTERRUPT_EVENT_COUNTER 1005
12)  /****************************/
13)  void InterruptTest(TestCase)
14)  {
15)     int fh;
16)     char bfr[82];
17)     struct timeval *tv;
18)     int *irqCount, *PulseWidth;
19)     printf("Interrupt test begin\n");
20)     fh = open("/dev/pport",O_RDWR);
21)     if (fh)
22)     {
```

```
23)        /* initialize bfr with all 0s */
24)        memset (bfr,'\0',82);
25)        switch(TestCase)
26)        {
27)        /* Will go to sleep until Interrupt occures */
28)        case WAIT_ON_INTERRUPT:
29)            ioctl (fh, WAIT_ON_INTERRUPT ,bfr);
30)            tv = (struct timeval *) bfr;
31)            printf("Got interrupt at %08u.%06u\n", (int)(tv->tv_sec %
               100000000),
32)            (int)(tv->tv_usec));
33)            break;
34)        /* If Interrupt has occurred then return the timeval else 0 */
35)        case INTERRUPT_PULSE_WIDTH:
36)           do
37)           {
38)              ioctl (fh, INTERRUPT_PULSE_WIDTH ,bfr);
39)              PulseWidth = (int *) bfr;
40)              sleep(1);
41)              printf("No Interrupt yet \n");
42)           } while(!(*PulseWidth));
43)           do
44)           {
45)               printf("Pulse Width=%d\n",*PulseWidth);
46)               ioctl (fh, INTERRUPT_PULSE_WIDTH ,bfr);
47)               PulseWidth = (int *) bfr;
48)           } while(*PulseWidth);
49)           break;
50)        case INTERRUPT_EVENT_COUNTER:
51)           do
52)           {
53)              printf("No Interrupt yet \n");
54)              sleep(1);
55)              ioctl (fh, INTERRUPT_EVENT_COUNTER ,bfr);
56)              irqCount = (int *) bfr;
57)           } while(!(*irqCount));
58)           printf("Interrupt Count = %d\n",*irqCount);
59)           break;
60)        default:
61)        }
62)        close(fh);
63)     }
64)     else
65)         printf("Fail to open device\n");
```

```
66) }
67) /*****************************/
68) int main(int argc , char *argv[])
69) {
70)     int pattern;
71)     while(1)
72)     {
73)        printf("*****Parallel port device driver Interrupt Test***\n");
74)        printf("Momentarily Connect parallel port pin10 to pin9\n");
75)        printf("Press Ctrl C to exit\n");
76)        printf("0 -- Test Sleep on Interrupt\n");
77)        printf("1 -- Test Interrupt Pulse Width\n");
78)        printf("2 -- Test Interrupt Count\n\n");
79)        printf("Enter Test #");
80)        scanf ("%x",&pattern);
81)        switch(pattern)
82)        {
83)        case 0: InterruptTest(WAIT_ON_INTERRUPT);
84)           break;
85)        case 1: InterruptTest(INTERRUPT_PULSE_WIDTH);
86)           break;
87)        case 2: InterruptTest(INTERRUPT_EVENT_COUNTER);
88)           break;
89)        }
90)     }
91) }
```

8. Edit the Makefile file /dd/interrupt/test/Makefile as shown in Listing 3.7, or copy it from the CD-ROM directory /book/chapter3/interrupt/test/Makefile. Compile and link the test.

```
#make
```

9. Change the mode of the output file t to make it executable.

```
# chmod 644 /dd/ioctl/test/t
```

Expected Test Results

The program will request a byte pattern to be written to the data register. In return, a byte pattern received from the status register will be displayed. If the loopback is correct, you should see the same pattern displayed on the screen.

Test Procedure

1. From the command line, execute the program t.

```
# ./t
```

2. Enter 0 to perform Sleep on Interupt. The routine will return only when you connect pins 10 and 9 to generate interrupt.
3. Enter 1 to receive the interrupt pulse width. The routine will display the time between interrupts.
4. Enter 2 to receive the interrupt counts.

Test Cleanup

ON THE CD

Press Ctrl-C to exit the program, and unload the device driver with unpp_load script as defined in Listing 3.4. The script is also available from the CD-ROM directory /book/chapter3/interrupt/pp_unload.

```
#  /dd/ioctl/pp_unload
```

Adding Task Scheduling to the Device Driver

ON THE CD

Quite often, a task in a device driver has to be executed periodically; for example, checking the key bounce on the keyboard, or monitoring the status of external signals in real time. In such cases, tasks are queued on the tq_timer task list using the queue_task function. The device driver creates a task during an init_module function call. The task performs the designated function and then puts itself back on the tq_timer task list, so that it can be executed periodically. It is not always possible to unregister such a device driver when the cleanup_module() is being called; for example, if the driver's task is queued up in the kernel's pending tasks list. One way to design a fail-safe mechanism is to let the cleanup module sleep_on an interrupt that is awakened when the task is no longer in the queue. The following is an explanation of the additions that we have performed for task scheduling capability to the device driver software of Listing 3.12. The file is available on the companion CD directory path /book/chapter3/task/pport.c.

- Line 30 is the wait_queue structure that is being initialized by the cleanup_module (line 284) as a signal to the pp_timer_interrupt task to abort the queue process (line 85).
- Lines 42 through 48 tq_struct Task initialize with the function pointer pp_timer_interrupt.

- Lines 61 through 92 are the function definition for the pp_timer_interrupt. This function will be executed periodically at every time tick. The current implementation simulates a key press condition on the status register of the parallel port. Any change in the state of the status port pins are being acknowledged as a key press event. At the end of the process, the function checks for a request to exit; if the WaitQ variable is non-NULL, the task will not be placed on the timer list (lines 85 through 92).
- Line 271 inserts the task at driver initialization time.
- Line 284 makes sure the function does not exit until all tasks are off the time list.

LISTING 3.12 File /book/chapter3/task/pport.c; a device driver with a task in the queue.

```
/*
* pport.c – Simple Hardware Operations and Raw Tests
*********/

1)    #include <linux/module.h>
2)    #include <linux/sched.h>
3)    #include <linux/wait.h>
4)    #include <linux/kernel.h> /* printk() */
5)    #include <linux/fs.h>     /* everything... */
6)    #include <linux/errno.h>  /* error codes */
7)    #include <linux/malloc.h>
8)    #include <linux/mm.h>
9)    #include <linux/ioport.h>
10)   #include <linux/interrupt.h>
11)   #include <linux/tqueue.h>
12)   #include <linux/time.h>
13)   #include <asm/io.h>
14)   #include <asm/segment.h>
15)   #include <linux/poll.h>
16)   #define CTRL_PORT_WR 1001
17)   #define CTRL_PORT_RD 1002
18)   #define WAIT_ON_INTERRUPT 1003
19)   #define INTERRUPT_PULSE_WIDTH 1004
20)   #define INTERRUPT_EVENT_COUNTER 1005
21)   #define READ_KEYS 1006
22)   static int pp_major = 0;
23)   static int pp_base = 0x378; /* intel: default to lp0 */
24)   static int pp_irq = 7;
25)   static unsigned long pp_buffer = 0;
```

```
26)  static int minor = 0;
27)  static int irqCount = 0;
28)  static unsigned long volatile pp_head;
29)  static volatile unsigned long pp_tail;
30)  static wait_queue_head_t pp_queue, WaitQ;
31)  static struct timeval current_time;
32)  static int TimerIntrpt=0;
33)  static int KeyPressed = 0;
34)  static int NewKey = 0;
35)  static int OriginalKey = 0;
36)  static int KeyAcknowledged = 0;
37)  /*
38)  The cleanup module uses this variable to find out if it is safe to unload
39)  */
40)  static void pp_timer_interpt (void *);
41)  /* The task queue structure for this task */
42)  static struct tq_struct Task =
43)  {
44)     {NULL,NULL},              /*Next item in the list */
45)     0,                 /* A flag to indicate task status */
46)     pp_timer_interpt,     /* The routine to be called */
47)     NULL           /* The parameter for the routine */
48)  };
49)  /*************** The pp_timer_interpt Task *****************/
50)  /*
51)  This function will be executed periodically at every timer tick
52)  as long as the device driver is active
53)  Simulate the key pressed condition using status register.
54)  Scan for status register and if there is any change then set the
55)  KeyPressed Variable and store the status register in NewKeys variable.
56)  The ioctl READ_KEYS will return the variable NewKeys and set
57)  KeyAcknowledge variable.
58)  The periodic scan will begin when status register has been returned
59)  to the original position
60)  */
61)  static void pp_timer_interpt (void * tick)
62)  {
63)     /* change the following code to suit your need */
64)     if (TimerIntrpt++ & 0x100)
65)     {
66)        /* every 100 tick read the status register */
67)        if (!KeyPressed)
68)        {
69)           NewKey = inb(pp_base+1) & 0xff;
```

```
70)             if (NewKey != OriginalKey)
71)             KeyPressed = 1;
72)          }
73)          else
74)          {
75)             if (KeyAcknowledged)
76)             {
77)                if((inb(pp_base+1) & 0xff) == OriginalKey)
78)                {
79)                   KeyPressed = 0;
80)                   KeyAcknowledged = 0;
81)                }
82)             }
83)          }
84)       }
85)       if (WaitQ.task_list.next != NULL)
86)       {
87)          printk(KERN_INFO "pp: Got to wake up ..\n");
88)          wake_up(&WaitQ);
89)       }
90)       else
91)          queue_task(&Task, &tq_timer);
92)    }
93)
94)    int pp_open (struct inode *inode, struct file *filp)
95)    {
96)       MOD_INC_USE_COUNT;
97)       minor =(MINOR(inode->i_rdev)&0x0f);
98)       printk(KERN_INFO "pp:U OPEN %d\n", minor);
99)       return 0;
100)   }
101)   int pp_close (struct file *filp)
102)   {
103)      MOD_DEC_USE_COUNT;
104)      return(0);
105)   }
106)   int pp_read (struct file * filp, char *buf, size_t count, loff_t * t)
107)   {
108)      unsigned char *kbuf=kmalloc(16, GFP_KERNEL);
109)      kbuf[1]='\0';
110)      kbuf[0] = inb(pp_base+1);
111)      /*    printk(KERN_INFO " READ VALUE pp: %x\n",*kbuf); */
112)      copy_to_user(buf, kbuf, 1);
113)      kfree(kbuf);
```

```
114)    return (0);
115) }
116) int pp_write (struct file * filp, const char *buf, size_t count, loff_t * t)
117) {
118)    int retval = count;
119)    unsigned char *kbuf=kmalloc(16, GFP_KERNEL);
120)    if (!kbuf)
121)       return -ENOMEM;
122)    copy_from_user(kbuf, buf, 1);
123)    /* printk(KERN_INFO "pp: WRITE  %x, %x\n",kbuf[0], kbuf[1]);  */
124)    outb(kbuf[0], pp_base);
125)    kfree(kbuf);
126)    return retval;
127) }
128) int pp_ioctl (struct inode *Inode, struct file *filp, unsigned int cmd,
     unsigned long arg)
129) {
130)    struct timeval tv;
131)    int key = 0;
132)    unsigned char *kbuf=kmalloc(16, GFP_KERNEL);
133)    if (!kbuf)
134)       return -ENOMEM;
135)    switch(cmd)
136)    {
137)    case READ_KEYS:
138)    /*
139)       Return any change in the status register.
140)       See pp_timer_interpt Task for details
141)    */
142)       if ((!KeyAcknowledged) && KeyPressed)
143)       {
144)          KeyAcknowledged = 1;
145)          switch(NewKey)
146)          {
147)          case 0x77: key = 1; /*status 3 pin 15 */
148)              break;
149)          case 0x6f: key = 2; /*status 2 pin 13 */
150)              break;
151)          case 0x5f: key = 3; /*status 5 pin 12 */
152)              break;
153)          case 0x3f: key = 4; /*status 4 pin 10 */
154)              break;
155)          case 0xff: key = 5; /*status 7 pin 11 */
156)              break;
```

```
157)          }
158)          copy_to_user((char *) arg, (char *) &key, sizeof (long));
159)       }
160)       break;
161)    case INTERRUPT_PULSE_WIDTH:
162)       /* interrupt pulse width as calculated by irq service routine */
163)       if (pp_head != pp_tail)
164)       {
165)          copy_to_user((char *) arg, (char *) pp_tail, sizeof (unsigned
          int));
166)          pp_tail = pp_tail + sizeof(unsigned int);
167)          if (pp_tail >= pp_buffer + PAGE_SIZE)
168)             pp_tail = pp_buffer; /* wrap around */
169)       }
170)       break;
171)    case INTERRUPT_EVENT_COUNTER:
172)       /* interrupt counts are being copied to arg pointer */
173)       copy_to_user((char *) arg, (char *) &irqCount, sizeof (long));
174)       irqCount = 0;
175)       break;
176)    case WAIT_ON_INTERRUPT:
177)       /* wait until an event occurs */
178)       init_waitqueue_head  (&pp_queue);
179)       interruptible_sleep_on(&pp_queue);
180)       /* fill in the time interrupt occurred */
181)       do_gettimeofday(&tv);
182)       copy_to_user((char *) arg, (char *) &tv, sizeof (struct timeval));
183)       break;
184)    case CTRL_PORT_WR:
185)       copy_from_user(kbuf, (char *) arg, 1);
186)       printk(KERN_INFO "pp: CTRL_PORT_WR  %x\n",kbuf[0]&0xff);
187)       outb(kbuf[0],pp_base+2);
188)       break;
189)    case CTRL_PORT_RD:
190)       kbuf[0] = inb(pp_base+2);
191)       printk(KERN_INFO "pp: CTRL_PORT_RD  %x\n",kbuf[0]&0xff);
192)       copy_to_user((char *) arg, kbuf, 1);
193)       break;
194)    default:
195)    }
196)    kfree(kbuf);
197)    return 0;
198) }
199) struct file_operations pp_fops = {
```

```
200) NULL,
201) NULL,          /* pp_lseek */
202) pp_read,
203) pp_write,
204) NULL,          /* pp_readdir */
205) NULL,          /* poll */
206) pp_ioctl,      /* ioctl */
207) NULL,          /* mmap */
208) pp_open,
209) pp_close,
210) NULL,          /* pp_fsync */
211) NULL,          /* pp_fasync */
212) /* nothing more, fill with NULLs */
213) };
214) void pp_interrupt(int irq, void *dev_id, struct pt_regs *regs)
215) {
216)    struct timeval tv;
217)    int usec, sec;
218)    unsigned * timediff;
219)    do_gettimeofday(&tv);
220)    /* calculate micro second difference */
221)    usec = tv.tv_usec - current_time.tv_usec;
222)    sec = tv.tv_sec - current_time.tv_sec;
223)    if (usec < 0)
224)        usec += 1000000; /* adjusted for rollover */
225)    current_time.tv_usec = tv.tv_usec;
226)    current_time.tv_sec  = tv.tv_sec;
227)    timediff = (unsigned *) pp_head;
228)    timediff = (sec * 1000000) + usec; /* save pulse width in usec */
229)    pp_head = pp_head + sizeof(unsigned int);
230)    if (pp_head == pp_buffer + PAGE_SIZE)
231)        pp_head = pp_buffer; /* wrap */
232)    irqCount++;
233)    printk(KERN_INFO "pp: INTERPT  %d\n",irqCount);
234)    wake_up_interruptible(&pp_queue);    /* awake any reading process */
235) }
236) int init_module(void)
237) {
238)    int result = check_region(pp_base,4);
239)    printk(KERN_INFO "pp: INIT_MOD\n");
240)    if (result)
241)    {
242)       printk(KERN_INFO "pp: can't get I/O address 0x%x\n",pp_base);
243)       return result;
```

```
244)     }
245)     request_region(pp_base,4,"pport");
246)     result = register_chrdev(pp_major, "pport", &pp_fops);
247)     if (result < 0)
248)     {
249)        printk(KERN_INFO "pp: can't get major number\n");
250)        release_region(pp_base,4);
251)        return result;
252)     }
253)     if (pp_major == 0) pp_major = result; /* dynamic */
254)     pp_buffer = __get_free_page(GFP_KERNEL); /* never fails */
255)     pp_head = pp_tail = pp_buffer;
256)     result = request_irq(pp_irq, pp_interrupt,
257)          SA_INTERRUPT, "pport", NULL);
258)     if (result)
259)     {
260)        printk(KERN_INFO "pp: can't get assigned irq %i\n", pp_irq);
261)        pp_irq = -1;
262)     }
263)     else
264)     {  /* actually enable it - assume this *is* a parallel port */
265)        printk(KERN_INFO "INTERRUPT ON %i\n", pp_irq);
266)        outb(0x10,pp_base+2);
267)     }
268)     WaitQ.task_list.next = NULL;
269)     init_waitqueue_head  (&pp_queue);
270)     do_gettimeofday(&current_time);
271)     queue_task (&Task, &tq_timer);
272)     OriginalKey =  inb(pp_base+1) & 0xff;
273)     return 0;
274) }
275) void cleanup_module(void)
276) {
277)     free_irq(pp_irq, NULL);
278)     unregister_chrdev(pp_major, "pport");
279)     release_region(pp_base,4);
280)     if (pp_buffer)
281)        free_page(pp_buffer);
282)     init_waitqueue_head  (&WaitQ);
283)     printk(KERN_INFO "pp: Wake me up ..\n");
284)     sleep_on(&WaitQ); /* exit only when task are free */
285)     printk(KERN_INFO "pp: Shutting down ..\n");
286) }
```

Testing the Task Scheduling

In this test scenario, the task scheduling of the pport device driver is being tested.

Test Objective

The objective of this test is to test the pport device driver's task scheduling implementation that simulates a key press condition with the help of the status port pins.

Test Description

This will test the pport device driver's ability to perform task scheduling at predefined intervals, and remove the task scheduling when the driver is unloaded. The test program displays any change on the status port pins as a key press event. Connecting pin-3 to pin-15 is shown as key#1, connecting pin-3 to pin-13 is shown as key#2, and so on—see the banner for a complete key map when you execute the test software.

Test Preparation

Figure 3.2 shows the pin-outs of the parallel port connector. The key map is displayed as follows

Connect parallel port pin-2 to pin-15 for key#1
Connect parallel port pin-2 to pin-13 for key#2
Connect parallel port pin-2 to pin-12 for key#3
Connect parallel port pin-2 to pin-10 for key#4
Connect parallel port pin-2 to pin-11 for key#5

Test Input

1. Mount the CD-ROM if you are planning to copy the source code from the CD.

```
# mount -t iso9660 /de/cdrom /mnt/cdrom
```

2. Create a subdirectory /dd/task, and edit the device driver source file /dd/task/pport.c as defined in Listing 3.12, or copy the source file from the CD-ROM directory /book/chapter3/task/pport.c.

```
# mkdir /dd/task
# cp /mnt/cdrom/book/chapter3/task/pport.c  /dd/task/
```

ON THE CD

3. Edit the device driver makefile /dd/task/Makefile as defined in Listing 3.3, or copy the source file from the CD-ROM directory /book/chapter3/task/Makefile.

```
# cp /mnt/cdrom/book/chapter3/task/Makefile  /dd/task/.
```

4. Compile and link the pport.c using the make utility.

```
# /dd/task/make
```

ON THE CD

5. Edit the script to load the device driver as defined in Listing 3.3, or copy it from the CD-ROM directory /book/chapter3/task/pp_load.

```
# cp /mnt/cdrom/dd/task/pp_load  /dd/task/.
```

6. Load the device driver using the pp_load script.

```
# /dd/task/pp_load
```

ON THE CD

7. Create a subdirectory /dd/task/test, and edit the file /dd/task/test/test.c as shown in Listing 3.13, or copy it from the CD-ROM directory /book/chapter3/task/test/test.c. The test program is a forever loop structure. It reads the interrupt status, and the timestamp of the interrupt is displayed.

```
# cp /mnt/cdrom/book/chapter3/task/test/*  /dd/task/test/.
```

LISTING 3.13 File /book/chapter3/task/test/test.c to test the pport device driver ioctl facility.

```
1   #include <stdio.h>
2)  #include <stdlib.h>
3)  #include <fcntl.h>
4)  #include <unistd.h>
5)  #include <string.h>
6)  #include <sys/ioctl.h>
7)  #define CTRL_PORT_WR 1001
8)  #define CTRL_PORT_RD 1002
9)  #define READ_KEYS 1006
10) /*****************************/
11) void KeyPressedTest(TestCase)
12) {
13)    int fh;
14)    char bfr[82];
```

```
15)    int *Keys;
16)    printf("Key press test begin\n");
17)    fh = open("/dev/pport",O_RDWR);
18)    if (fh)
19)    {
20)       /* initialize bfr with all 0s */
21)       /* disable interrupt */
22)       ioctl (fh, CTRL_PORT_WR,0);
23)       switch(TestCase)
24)       {
25)       case READ_KEYS:
26)          do
27)          {
28)             memset (bfr,'\0',82);
29)             ioctl (fh, READ_KEYS ,bfr);
30)             Keys = (int *) bfr;
31)             sleep(1);
32)             printf("No Keys yet \n");
33)             if (*(Keys))
34)                 printf("Key pressed=%x\n",*Keys);
35)          } while(1);
36)          break;
37)       default:
38)       }
39)       close(fh);
40)    }
41)    else
42)       printf("Fail to open device\n");
43) }
44) /*****************************/
45) int main(int argc , char *argv[])
46) {
47)    while(1)
48)    {
49)       printf("*****Parallel Port Device Key Scan Task Test***\n");
50)       printf("Connect parallel port pin-2 to pin-15 for key#1\n");
51)       printf("Connect parallel port pin-2 to pin-13 for key#2\n");
52)       printf("Connect parallel port pin-2 to pin-12 for key#3\n");
53)       printf("Connect parallel port pin-2 to pin-10 for key#4\n");
54)       printf("Connect parallel port pin-2 to pin-11 for key#5\n");
55)       printf("Press Ctrl C to exit\n");
56)       KeyPressedTest(READ_KEYS);
57)    }
58) }
```

ON THE CD

8. Edit the Makefile file /dd/task/test/Makefile as shown in Listing 3.7, or copy it from the CD-ROM directory /book/chapter3/task/test/Makefile. Compile and link the test.

```
#make
```

9. Change the mode of the output file t to make it executable.

```
# chmod 644 /dd/task/test/t
```

Expected Test Results

The program displays the banner and waits for a key to be pressed. Pressing a key momentarily will be acknowledged by displaying it on the screen. Pressing Ctrl-C will abort the program and unload the task that was being queued up for scheduling.

Test Procedure

1. From the command line, execute the program t.

```
# ./t
```

2. The program will wait for a key press.

Test Cleanup

ON THE CD

Press Ctrl-C to exit the program, and unload the device driver with pp_unload script as defined in Listing 3.4. The script is also available from the CD-ROM directory /book/chapter3/task/pp_unload.

```
[root@ns /dd/task]#  /dd/task/pp_unload
```

SUMMARY

Device drivers are special programs that run in the kernel memory space performing low-level I/O functions on the hardware devices attached to the system. The kernel assigns each device a node number just like it was a file in the system, and issues a name during initialization when the device is loaded into the system. The /dev directory is generally reserved for the device files, but their names also appear in the /proc/devices directory in the Linux operating system. The device drivers are created as either char device drivers or as block device drivers; the difference is the access method, but they appear similar from a functional point of view.

Making device files is a two-step process:

1. Insert the module with the insmod command; the kernel then places the major number into the /proc/devices file.
2. Attach a filename with a file node with the mknod command.

We covered the essentials of device driver design using parallel port as an example.

4 Tasks and Interprocess Communication

In this chapter

- Tasks
- Interprocess Communication Facilities
- Shared Data Structures
- Semaphores
- Message Queues
- Pipes
- Fork
- Sockets

INTRODUCTION

Computer programming is about processing information and presenting the result as desired by the user. For that reason, we have divided programming into the *system interface* and the *user interface*. The system interface is where the acquired data are processed and manipulated to achieve the desired result. Presenting the result to the user is the job of the user interface. The Internet is the media of choice for user interface, and in Chapter 2, "The User Interface," you saw how easy it is to use the Linux tools for developing a robust Internet solution. The system interface, on the other hand, can be an involved process and might require interface between several modules as the processing of data goes through different stages of completion. It is important to establish a client/server relationship between modules and set up a communication protocol among them so that each module can be treated as a separate replaceable entity. Remember, modules can come and go, but protocols are here to stay.

You saw the first part of the system interface design in Chapter 3, "Device Drivers," that showed how data are acquired through external sources. In this chapter, we will discuss the client/server portion of the system interface design that involves processing of data as it goes through different modules. The operating system plays an important role in providing services that establish communication between modules, and this chapter is devoted to the specific services that the Linux operating system provides for establishing an efficient client/server mechanism.

An operating system is basically a set of routines that work closely with the hardware platform of the system that acts as its foundation. It also provides an environment for the application programs to access the underlying platform hardware, such as the video card, the keyboard, the mouse, the network card, and so forth. One important feature common in most modern real-time operating systems is the facility for simultaneously executing several small programs. It gives the illusion that at any given time, a single task or program has all the system resources under its control. Unix had this feature available from its inception. The importance of this feature was recognized from the early days of computer development, and the multitasking kernel was considered essential for any decent software application.

A multitasking kernel allows you to organize your software into smaller, manageable tasks. For example, a clear division of labor in a real-time data acquisition system is acquiring the data, processing the data, displaying the information to the user, acknowledging interrupts from the user, alarm and error reporting, and so forth. One monolithic sequential program will be a disaster in this situation. What we need is an independent execution of smaller tasks that are completely enclosed and oblivious to the fact that there are other pieces of software requiring attention. This does not mean that a task should unnecessarily stay in a loop doing nothing. Real-time multitasking kernels provide proper procedures for tasks to relinquish CPU resources at the appropriate time. In a true multitasking environment, the kernel knows how to divide the CPU time judiciously among the tasks that are present in the system. Another key facility in real-time systems is hardware interrupt handling, as the interrupts are the usual mechanism to inform a system of external events.

TASKS

The building blocks of computer program applications are the set of modules developed to perform specified tasks. A multitasking environment allows a real-time application to be constructed as a set of independent *tasks*, each with its own thread of execution, while keeping a set of system resources at their disposal. When several of these modules are linked together and executed in computer system memory, they all compete for the same system resources. Tasks also need some way to synchronize events among themselves, and require some way of channeling parameters and messages to other routine or routines, while receiving messages and parameters from others at the same time. If not designed carefully, tasks can become "deadlocked"; task A waits for resources from task B, while task B waits for resources from task A, thereby keeping each other in limbo (Figure 4.1).

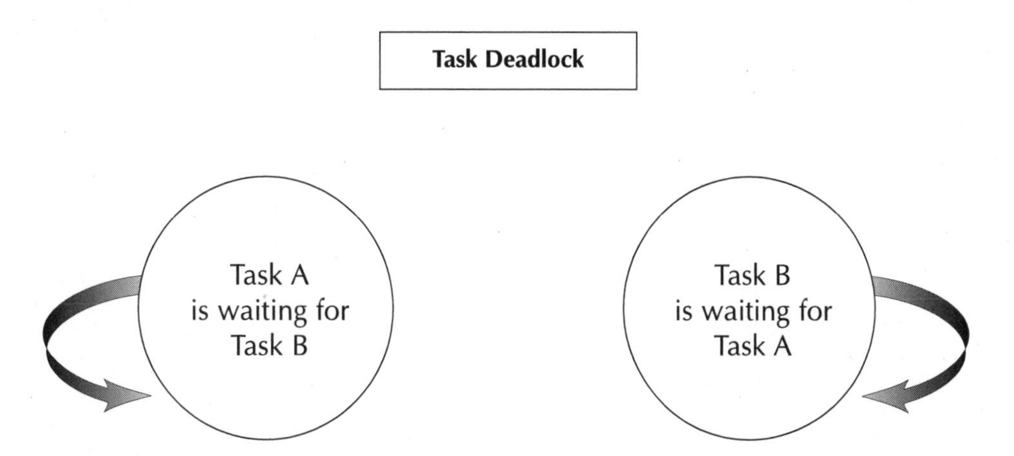

FIGURE 4.1 Deadlocks occur when tasks wait for resources that were not released at the proper time.

It is quite possible to design mechanisms in software by the application programs themselves that are capable of handling the requirements of event synchronization and passing parameters. However, that would be a waste of time. An operating system such as Linux has well-defined and robust mechanisms in place that can handle all such interprocess communication requirements, even in extremely complex situations.

The interprocess communication facilities of Linux allow tasks and processes to synchronize and communicate with each other, and coordinate the activities such as waiting for messages, passing parameters, sharing system resources, and so forth. The key interprocess communication facilities are *shared memories*, *semaphores*, *message queues*, *pipes*, *network transparent sockets*, and *signals*.

A set of standards known as POSIX specifies the interfaces between the methods of the intertask communication facilities in order to facilitate a coherency and interoperability across different platforms and operating systems. The POSIX standard (1003.1b) defines a set of interfaces to kernel facilities to improve application portability, and is supported by several operating system vendors such as VxWorks and QMS. The Linux kernel differs slightly from the POSIX standard in some respect, although it has a rich set of facilities for interprocess communication.

A *task* might be defined as a thread of execution running independently in a real-time operating system such as Linux. It is often essential to organize applications into independent processes or tasks so that the software design of the program remains within a manageable boundary. In an operating system paradigm, a task is a kernel data structure that holds information about a process and the system call it is in. Tasks have immediate, shared access to most system resources, yet maintain enough separate context to maintain an individual thread of control.

Multitasking

Each application program is initially dispatched as a single task of execution by the operating system at the time of its launch by the user or by the system. The program might spawn other tasks of its own once it starts running; thus, at any given time, several tasks are in the state of execution. The operating system simply maintains the schedule and the system resources that these tasks occupy. A *multitasking kernel* is capable of scheduling multiple tasks so that each has sufficient system resources at a specified time.

Preemptive multitasking forces tasks to give up system resources after a specified amount of time. Linux is a preemptive multitasking kernel. It is possible to have the appearance of multitasking by using cooperative programs that release system resources voluntarily so other programs waiting in line can resume execution. This process is known as *non-preemptive multitasking*. Multitasking provides the fundamental mechanism for an application to control and react to multiple, discrete real-world events. Each task has its own context, which is the CPU environment and system resources belonging to the task. On a context switch, a task's context is saved in the task control block (TCB), and the task next in line is awakened and executed. A task's context includes:

- A thread of execution; that is, the task's program counter.
- The CPU registers.
- A stack for dynamic variables and function calls.
- I/O assignments for standard input, output, and error.
- A delay timer.
- Kernel control structures.
- Signal handlers.

Multitasking versus Multiprogramming

An operating system capable of simultaneously executing multiple programs is also a multitasking system, but there is a significant difference between several tasks running in an application versus several processes running in an operating system. All processes own individual memory space; one process cannot access the memory space of another (except when memory sharing). Each process inherits many system resources, such as input and output stream, environment variables, permission setting, and so forth. These resources can only be shared among its child processes, whereas multiple tasks share the same memory and resources and run inside a single process.

Implementing Multitasking in Linux

Support of POSIX standards for multitasking is deficient in Linux (things might change by the time you read this book), but implementing a multitasking program in Linux is not difficult. In fact, a rich set of interprocess communication facilities gives you a very good foundation on which to build complex multitasking programs. In this section, we will build a program with three different tasks to demonstrate the framework for designing multitasking programs (Listing 4.2). The limitation of this program is that it requires the kernel's memory space for execution. Each task runs through completion and puts itself back to sleep. It is a simple application, but later we will add the interprocess communication facilities to demonstrate the techniques of task synchronization and resource sharing.

Asking the Linux kernel to call a specified function every time the timer ticks is a simple process. Define a struct tq_struct that holds a pointer to the function, and call the kernel function queue_task(), passing the parameter tq_struct. The kernel will add the task to its task list, and the next time the timer ticks, the function specified in the tq_struct will be executed. Once the function is called, the kernel will remove the tq_struct from the list. It is the task function's responsibility to put itself back in the queue if it wants to be called again. Unfortunately, there is a slight problem with this scheme. If someone wants to remove the module by pressing Ctrl-C, there is still a reference to the function in the task queue. Therefore, when the kernel calls the function at the next timer tick, the function is no longer in place. To avoid this situation, the signal handler is modified so that instead of exiting immediately, it informs the task function that the job is complete and it does not need to be put back on the task list.

ON THE CD

Listing 4.1 is a simple multitasking program with three tasks created by the init_module function. The source code is available in the companion CD directory /book/chapter4/multitasking/task.c. The important thing to remember is that before exiting the process, we need to make sure that tasks are not left on the task queue list; otherwise, the kernel will find a pending task pointer that does not exist in the system.

ON THE CD

Use the Makefile provided in Listing 4.2 to create the module. The Makefile is also available in the companion CD directory /book/chapter4/multitasking/Makefile.

Execute the program using the following shell command:

```
# insmod -f tasks.o tasks
```

Verify the execution of tasks by monitoring the output of the kernel log using the following shell command:

```
# cat /var/log/messages
```

You would see the following messages scroll by indicating the output from the task:

```
Oct 31 08:31:54 ns kernel: Task B is in foreground after 1000 ticks
Oct 31 08:32:04 ns kernel: Task B is in foreground after 1000 ticks
Oct 31 08:32:14 ns kernel: Task A is in foreground after 3000 ticks
Oct 31 08:32:14 ns kernel: Task B is in foreground after 1000 ticks
Oct 31 08:32:24 ns kernel: Task C is in foreground after 4000 ticks
Oct 31 08:32:24 ns kernel: Task B is in foreground after 1000 ticks
Oct 31 08:32:34 ns kernel: Task B is in foreground after 1000 ticks
Oct 31 08:32:44 ns kernel: Task A is in foreground after 3000 ticks
```

Remove the module using the following shell command:

```
# rmmod tasks
```

LISTING 4.1 File /book/chater4/multitasking/task.c; a framework for multitasking implementation.

```
1)    #include <linux/module.h>
2)    #include <linux/sched.h>
3)    #include <linux/wait.h>
4)    #include <linux/kernel.h> /* printk() */
5)    #include <linux/fs.h>     /* everything... */
6)    #include <linux/errno.h>  /* error codes */
7)    #include <linux/malloc.h>
8)    #include <linux/mm.h>
9)    #include <linux/ioport.h>
10)   #include <linux/interrupt.h>
```

```
11)  #include <linux/tqueue.h>
12)  #include <linux/time.h>
13)  #include <linux/signal.h>
14)  #include <asm/io.h>
15)  #include <asm/segment.h>
16)  #include <linux/poll.h>
17)  #include <stdio.h>
18)  static void Task_B (void *);
19)  static void Task_A (void *);
20)  static void Task_C (void *);
21)  static wait_queue_head_t TaskAWaitQ;
22)  static wait_queue_head_t TaskBWaitQ;
23)  static wait_queue_head_t TaskCWaitQ;
24)  static int TasksRunning = 1;
25)  /***********************************************/
26)  /*
27)  The task queue structure for task A
28)  */
29)  static struct tq_struct TaskAStruct =
30)  {
31)  {NULL,NULL},    /* Next item in the list */
32)  0,       /* A flag to indicate not being inserted yet */
33)  Task_A, /* The routine to be called */
34)  NULL     /* The parameter for the routine */
35)  };
36)  /***********************************************/
37)  /*
38)  The task queue structure for task B
39)  */
40)  static struct tq_struct TaskBStruct =
41)  {
42)  {NULL,NULL},    /* Next item in the list */
43)  0,       /* A flag to indicate not being inserted yet */
44)  Task_B, /* The routine to be called */
45)  NULL     /* The parameter for the routine */
46)  };
47)  /***********************************************/
48)  /*
49)  The task queue structure for task C
50)  */
51)  static struct tq_struct TaskCStruct =
52)  {
53)  {NULL,NULL},    /* Next item in the list */
54)  0,       /* A flag to indicate not being inserted yet */
```

```
55)  Task_C, /* The routine to be called */
56)  NULL    /* The parameter for the routine */
57)  };
58)  /****************** The Task A *******************/
59)  /*
60)  This function will be executed periodically at every timer tick
61)  as long as the device driver is active. Wakes up after 200 ticks
62)  */
63)  static void Task_A (void * tmp)
64)  {
65)     static int TimerIntrpt = 0;
66)     if (TimerIntrpt++ >= 3000)
67)     {
68)        // user process starts here
69)        TimerIntrpt = 0;
70)        printk(KERN_INFO "Task A is in foreground after 3000 ticks\n");
71)        // user process ends here
72)     }
73)     if (TaskAWaitQ.task_list.next != NULL)
74)        wake_up(&TaskAWaitQ);
75)     else
76)        queue_task(&TaskAStruct, &tq_timer);
77)  }
78)  /****************** The Task B *******************/
79)  /*
80)  This function will be executed periodically at every timer tick
81)  as long as the device driver is active. Wakes up after 200 ticks
82)  */
83)  static void Task_B (void * tmp)
84)  {
85)     static int TimerIntrpt = 0;
86)     if (TimerIntrpt++ >= 1000)
87)     {
88)        // user process starts here
89)        TimerIntrpt = 0;
90)        printk(KERN_INFO "Task B is in foreground after 1000 ticks\n");
91)        // user process ends here
92)     }
93)     if (TaskBWaitQ.task_list.next != NULL)
94)        wake_up(&TaskBWaitQ);
95)     else
96)        queue_task(&TaskBStruct, &tq_timer);
97)  }
98)  /****************** The Task C *******************/
```

```
99)  /*
100) This function will be executed periodically at every timer tick
101) as long as the device driver is active. Wakes up after 4000 ticks
102) */
103) static void Task_C (void * tmp)
104) {
105)    static int TimerIntrpt = 0;
106)    if (TimerIntrpt++ >= 4000)
107)    {
108)       // user process starts here
109)       TimerIntrpt = 0;
110)       printk(KERN_INFO "Task C is in foreground after 4000 ticks\n");
111)       // user process ends here
112)    }
113)    if (TaskCWaitQ.task_list.next != NULL)
114)        wake_up(&TaskCWaitQ);
115)    else
116)        queue_task(&TaskCStruct, &tq_timer);
117) }
118) /****************************** Startup  ******************************/
119) /* Start three tasks and set them free */
120) int init_module(void)
121) {
122)    TaskAWaitQ.task_list.next = NULL;
123)    TaskBWaitQ.task_list.next = NULL;
124)    TaskCWaitQ.task_list.next = NULL;
125)    queue_task (&TaskAStruct, &tq_timer);
126)    queue_task (&TaskBStruct, &tq_timer);
127)    queue_task (&TaskCStruct, &tq_timer);
128)    return 0;
129) }
130) /*************************** Shutdown **************************/
131) /*
132) The ending of life of the process, but first wait
133) for the task to be back from sleep
134) */
135) void cleanup_module(void)
136) {
137)    init_waitqueue_head  (&TaskAWaitQ);
138)    printk(KERN_INFO "UC: Task A closing down ..\n");
139)    sleep_on(&TaskAWaitQ); /* exit only when task are free */
140)    init_waitqueue_head  (&TaskBWaitQ);
141)    printk(KERN_INFO "UC: Task B closing down ..\n");
142)    sleep_on(&TaskBWaitQ); /* exit only when task are free */
```

```
143)      init_waitqueue_head  (&TaskCWaitQ);
144)      printk(KERN_INFO “UC: Task C closing down ..\n”);
145)      sleep_on(&TaskCWaitQ); /* exit only when task are free */
146)      TasksRunning = 0;
147)      printk(KERN_INFO “pp: Shutting down All Tasks ..\n”);
148)  }
```

LISTING 4.2 File /book/chater4/multitasking/Makefile; the Makefile to create task.o module.

```
1)  INCLUDEDIR = /usr/include
2)  DEBFLAGS = -02
3)  CFLAGS = -D__KERNEL__ -DMODULE -Wall $(DEBFLAGS)
4)  CFLAGS += -I$(INCLUDEDIR)
5)  OBJS = tasks.o
6)  all: $(OBJS)
```

INTERPROCESS COMMUNICATION FACILITIES

Processes in Linux do not have enough information about each other to communicate directly or pass parameters. They live in a virtual world of their own, where the paging mechanism of the operating system hides all the details of the memory that belongs to a specific process. Only the underlying operating system has the true knowledge of the physical environment that can make an effective link between the different processes and allow the parameters to be passed. To do so (pass parameters or share parameters among tasks and processes), operating systems expose certain function calls called *interprocess communication facilities.*

Tasks, on the other hand, can share variables, as they are executed in the same memory space; however, it is a dangerous practice to depend on global variables for task activities. Therefore, it is recommended that individual tasks be designed around the facilities of interprocess communication as if they are independent processes. Each mechanism of the facility provides a unique environment and is suitable for a specific need.

The following are Linux-provided interprocess communication methods:

- Shared memory, for simple sharing of data
- Semaphores, for basic mutual exclusion and synchronization
- Message queues
- Sockets and remote procedure calls, for network-transparent intertask communication

- Signals, for exception handling
- Fork and pipes

SHARED DATA STRUCTURES

The most obvious way for processes to communicate among each other is by accessing shared data structures. It is easier to share data in a single linear address space (as in multitasking); however, for virtual memory operating systems such as Linux, it is necessary to set aside a memory segment as shared and then allow processes to access the same segment. A segment can be created by one process and subsequently accessed for read and write by other processes. The concept of shared memory might not be applicable directly for a multitasking environment, as the tasks already run in the same physical memory space. However, tasks and processes are interchangeable, so they should follow the same rule as the processes do, which is what we will demonstrate in the coming examples.

The Linux kernel maintains a data structure shmid_ds for each shared memory segment as described in the following section.

```
1) struct shmid_ds
2) {
3)      struct ipc_perm shm_perm;/*permission */
4)      intshm_segsz;/* size of segments in bytes */
5)      time_tshm_atime;/* last attach time */
6)      time_tshm_dtime;/* last detach time */
7)      time_tshm_ctime;/* last change time */
8)      unsigned short shm_cpid;/* pid of creator */
9)      unsigned short shm_lpid;/* pid of last operator */
10)     shortshm_nattach;/* number of current attaches */
11) unsigned short shm_npages;/* size of segment pages */
12)     unsigned long *shm_pages; /* array of ptrs to frame */
13)     struct vm_area_struct attaches;/* descriptor for attach */
14) };
```

Shared Memory Support Functions

The Linux kernel provides the following system calls for application programs to create and manage the shared memory data structure:

shemget() Create a new shared memory region, or access an existing region.

shemat() Attach or map the segment into its own addressing space.

shemctl() Allows control of the data structure, such as changing permissions, etc.

shemdt() Detach a shared memory structure.

shmget

Create a new shared memory region, or access an existing region.

Prototype

```
int shmget( key_t key, int size, int shmflg);
```

Parameters

```
key_t key:
The kernel tries to match this with exiting key values, the open or
access operation depends upon the flag shmflg
    Size:
            The requested size of the segment
    shmflg:
            combination of the following types
            IPC_CREAT
Create the segment if does not exist already else return the existing
segment identifier
            IPC_EXCL
When used with IPC_CREAT, fails if segment already exist
```

Returns

```
shared memory segment identifier
or -1 on error
    errno
            = EINVL (invalid segment size)
            = EEXIST (Segment exist, cannot create)
            = EIDRM (segment reserved)
            = ENOENT (Segment does not exist)
            = EACCESS (Permission denied)
            = ENOMEM (Not enough memory)
```

shmat

```
    Attach or map the segment into its own addressing space
```

Prototype

```
int shmat( int shmid, char * shmaddr, int shmflg);
```

Parameters

```
    int shmid:
The id returned from the shmget
    Char * shmaddr:
The address to be used for mapping, the recommended value is zero so
the kernel will try to find an unmapped region.
    shmflg:
            The following types
            SHM_RND
Force a passed address to be page aligned
            SHM_RDONLY
                    Read only segment
```

Returns

```
Address at which segment was attached
or -1 on error
    errno
= EINVL (invalid segment size)
```

shmctl

Allows control of the data structure, such as changing permissions and removing shared memory segment allocation.

Prototype

```
int shmctl( int shmid, int cmd, struct shmid_ds * buf);
```

Parameters

```
    int shmid:
The id returned from the shmget
    int cmd:
            IPC_STAT
            Retrieves the shmid_ds structure for a segment and stores
            it in the address of the buf argument
```

```
IPC_SET
      Sets the value of the (permission) ipc_perm member of the
      shmid_dsstructure. The value is taken from the buf argument
IPC_RMID
      Marks a segment for removal
      Shmid_ds buf
            The buffer for the commands
```

Returns

```
shared memory segment identifier
or -1 on error
   errno
         = EACCESS (Permission denied)
         = EFAULT (Address pointed to by buf is invalid)
         = EIDRM (segment reserved)
         = EINVAL (Shmid is invalid)
         = EPERM (No write access to the segment)
```

shmdt

```
detach a shared memory structure
```

Prototype

```
int shmgdt( char * shmaddr);
```

Parameters

```
Char * shmaddr:
   The address used for mapping
```

Returns

```
Address at which segment was attached
or -1 on error
   errno
         = EINVL (invalid segment size)
```

Shared Memory Implementation

ON THE CD Listing 4.3 is an example of a shared memory support function implementation. The source code is available in the companion CD directory /book/chapter4/shmem/shmem.c. The program simply requests the kernel to create a shred memory structure (line 22) using the key SHARED_KEY (line 9). The shared memory pointer returned from the kernel is being mapped to a local variable, which is being used in subsequent operations. The program simply waits for other programs to communicate through its shared memory structure. Use the Makefile (provided in the companion CD directory /book/chapter4/shmem/Makefile) in Listing 4.4 to compile the program, and execute it using the following shell command:

```
# cd /book/chapter4/shmem
# make
# ./t
```

LISTING 4.3 File /book/chapter4/shmem/shmem.c; a shared memory implementation.

```
1)    #include <unistd.h>
2)    #include <stdlib.h>
3)    #include <stdio.h>
4)    #include <string.h>
5)    #include <sys/types.h>
6)    #include <sys/ipc.h>
7)    #include <sys/shm.h>
8)    #define MEM_SZ 4096
9)    #define SHARED_KEY 9988
10)   struct shared_struct
11)   {
12)   int flag;
13)   char bfr[80];
14)   };
15)   char * shared_memory;
16)   struct shared_struct * shared;
17)   struct shmid_ds shmds;
18)   int shmid;
19)   /************************/
20)   int OpenSharedMem()
21)   {
22)      shmid = shmget (SHARED_KEY, MEM_SZ, 0666 | IPC_EXCL | IPC_CREAT);
23)      if (shmid == -1)
24)      {
25)         shmid = shmget (SHARED_KEY, MEM_SZ, 0666 | IPC_CREAT);
26)         if (shmid == -1)
27)         {
28)            fprintf(stderr, "shmget failed \n");
29)            exit (EXIT_FAILURE);
30)         }
```

```
31)         else
32)         {
33)            printf("Open Shared memory as client\n");
34)         }
35)      }
36)      else
37)         printf("Shared memory created\n");
38)      shared_memory = shmat (shmid, (void *)0, 0);
39)      if (shared_memory == (void *) -1)
40)      {
41)         fprintf(stderr, "shmat failed \n");
42)         exit (EXIT_FAILURE);
43)      }
44)      printf("memory attached at %x\n", (int) shared_memory);
45)      shared = (struct shared_struct *) shared_memory;
46)      shared->flag = 0;
47)      shared->bfr[0] = '\0';
48)      return (0);
49)   }
50)   /************************/
51)   void DeleteSharedMem()
52)   {
53)      shmctl(shmid, IPC_RMID, 0);
54)      printf("Deleteing shared mem\n");
55)      sleep(1);
56)      OpenSharedMem();
57)   }
58)   /************************/
59)   void ModeDisplay()
60)   {
61)      shmctl (shmid, IPC_STAT, &shmds);
62)      printf("Shared mem permission %x\n",shmds.shm_perm.mode);
63)   }
64)   /************************/
65)   int main()
66)   {
67)      int running = 1;
68)      OpenSharedMem();
69)      while(running)
70)      {
71)         if (  shared->flag)
72)         {
73)            printf ("Msg in shared mem %d\n",  shared->flag);
74)            switch(shared->flag)
```

```
75)            {
76)            case 1: /* Read */
77)               printf ("Message from task B %s\n",shared->bfr);
78)               break;
79)            case 2: /* Write */
80)               strcpy(shared->bfr,"This is a Message from task A \n");;
81)               break;
82)            case 3: /* Delete */
83)               DeleteSharedMem();
84)               break;
85)            case 4: /* Mode display */
86)               shmctl (shmid, IPC_STAT, &shmds);
87)               printf("Shared mem permission %x\n",shmds.shm_perm.mode);
88)               break;
89)            }
90)            strcpy(shared->bfr,"Done \n");
91)            shared->flag = 0;
92)         }
93)         else
94)         {
95)             printf ("No msg yet \n");
96)         }
97)         sleep(1);
98)      }
99)      return 0;
100) }
```

LISTING 4.4 File /book/chapter4/shmem/Makefile; a Makefile to create the shmem.o executable.

```
all: shmem
t:CC = gcc
t:INCLUDEDIR = /usr/include
t:CFLAGS = -g -02 -Wall -I$(INCLUDEDIR)
t:shmem: shmem.o
    $(CC) -o t shmem.o
t:shmem.o: shmem.c
    $(CC) -c shmem.c $(CFLAGS)
```

ON THE CD

A test program is being developed in Listing 4.5 (you can copy the source from the CD directory /book/chpater4/shmem/test/test.c) that communicates with the original shmem module and passes parameters using the shared memory structure. The construct of the file is similar to the original program shmem.c and uses the same SHARED_KEY to create a shared memory structure. Compile the program using the Makefile in Listing 4.6 (see the companion CD directory /book/chapter4/shmem/test/Makefile). Execute the program using the following shell command:

```
# cd /book/chapter4/shmem/test
# make
# ./t
```

Upon execution of the test program, messages through shared memory are displayed on the screen.

LISTING 4.5 File /book/chapter4/shmem/test/test.c; a shared memory test implementation.

```
1)    #include <unistd.h>
2)    #include <stdlib.h>
3)    #include <stdio.h>
4)    #include <sys/types.h>
5)    #include <sys/ipc.h>
6)    #include <sys/shm.h>
7)    #define MEM_SZ 4096
8)    #define SHARED_KEY 9988
9)    int pattern;
10)   struct shared_struct
11)   {
12)   int flag;
13)   char bfr[80];
14)   };
15)   char * shared_memory;
16)   struct shared_struct * shared;
17)   int shmid;
18)   int OpenSharedMem()
19)   {
20)      shmid = shmget (SHARED_KEY, MEM_SZ, 0666 | IPC_EXCL | IPC_CREAT);
21)      if (shmid == -1)
22)      {
23)         shmid = shmget (SHARED_KEY, MEM_SZ, 0666 | IPC_CREAT);
24)         if (shmid == -1)
25)         {
26)            fprintf(stderr, "shmget failed \n");
27)            exit (EXIT_FAILURE);
28)         }
```

```
29)           else
30)           {
31)              printf("Open Shared memory as client\n");
32)           }
33)        }
34)        else
35)           printf("Shared memory created\n");
36)        shared_memory = shmat (shmid, (void *)0, 0);
37)        if (shared_memory == (void *) -1)
38)        {
39)           fprintf(stderr, "shmat failed \n");
40)           exit (EXIT_FAILURE);
41)        }
42)        printf("memory attached at %x\n", (int) shared_memory);
43)        shared = (struct shared_struct *) shared_memory;
44)        shared->flag = 0;
45)        return (0);
46)     }
47)     int main()
48)     {
49)        while(1)
50)        {
51)           OpenSharedMem();
52)           printf("**** Shared memory test ****\n");
53)           printf("Press Ctrl C to exit\n");
54)           printf("1 -- Test Shared Read\n");
55)           printf("2 -- Test Shared Write\n");
56)           printf("3 -- Test Shared Delete\n");
57)           printf("4 -- Test Shared Permission\n");
58)           printf("Enter Test #");
59)           scanf ("%x",&pattern);
60)           shared->flag = pattern;
61)           sleep(1);
62)           printf("%s",shared->bfr);
63)        }
64)     }
```

LISTING 4.6 File /book/chapter4/shmem/test/Makefile; a Makefile to create the test.o executable.

```
all: t
CC = gcc
```

```
INCLUDEDIR = /usr/include
CFLAGS = -g -O2 -Wall -I$(INCLUDEDIR)

t: test.o
	$(CC) -o t test.o

test.o: test.c
	$(CC) -c test.c $(CFLAGS)
```

Mutual Exclusion in Shared Memory

While a shared address space simplifies exchange of data, interlocking access to memory is crucial to avoid contention. Many methods exist for obtaining exclusive access to resources, and vary only in the scope of the exclusion. Such methods include disabling interrupts and disabling preemption, but the most desirable is the resource locking with semaphores, which we discuss next.

Semaphores

In a multitasking system, there is no guarantee as to when a task gives up its hold on a shared resource. For example, task A just read a shared variable whose value was 0, incremented the contents, but was switched out of context before it had a chance to write the value back into the memory. Task B wakes up, reads the contents of the shared variable, which is still a 0, decrements the contents, and writes back the new value into the memory. When task A resumes execution, it finishes its job, which is to write back the value 1 into the shared variable, thus nullifying the effect of task B—sheer disaster in computer programming. What we need is a mutual exclusion of tasks when a read/modify/write operation is taking place.

Semaphores are the primary means for addressing such requirements of mutual exclusion among tasks. For mutual exclusion, semaphores interlock access to shared resources. Semaphores are also used in task synchronization, as well as coordinating a task's execution with external events.

The concept of semaphore mechanism is rather novel, but very simple in implementation. A semaphore is basically a kernel resource that several tasks try to take possession of. The task that is granted the possession of the semaphore proceeds further to act upon the shared resource. Once the task is finished with the resource, it informs the kernel that the semaphore is no longer needed. The kernel then switches the current task out of context and allows the next task in line to start. The next task now has sole possession of the shared resource, and all other tasks wait for the kernel to grant them access to the semaphore. Figure 4.2 depicts the stages of task A and task B. When task A calls the get routine, it does not return, while task B has already passed the get state.

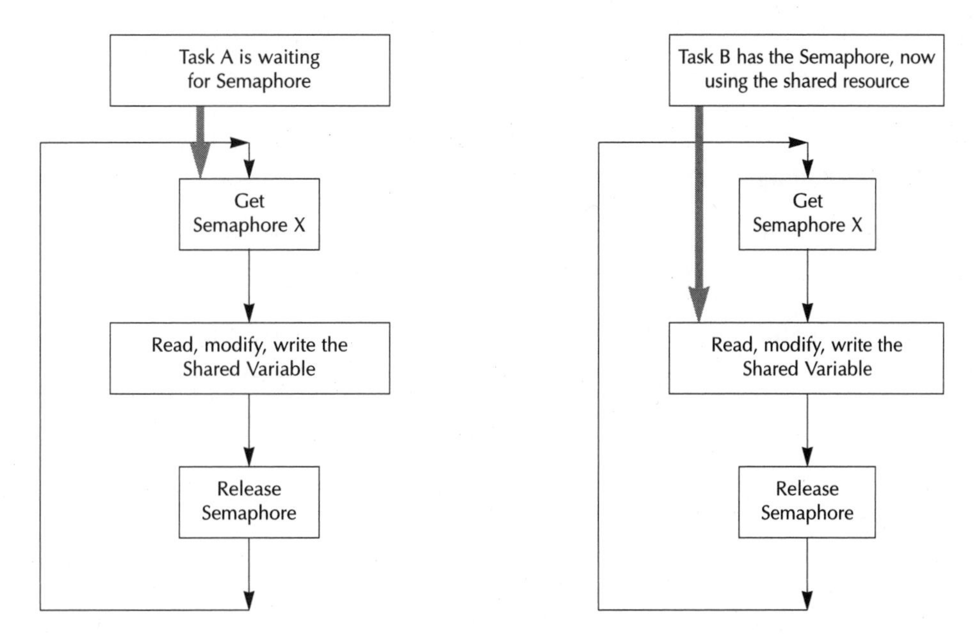

FIGURE 4.2 The state of different tasks while attempting to gain access to the same resource.

Similar to the shmid_ds structure of the shared memory, the kernel maintains a special internal data structure in the include file <sys/sem.h> semid_ds for semaphores.

```
1)  struct semid_ds
2)  {
3)     struct ipc_perm sem_perm;/*permission */
4)     time_tshm_otime;/*last sem operation time */
5)     time_tshm_ctime;/*last change time */
6)     struct sem*sem_base;/*ptr to first sem in array*/
7)     struct wait_queue *eventn;
8)     struct wait_queue *eventz;
9)     struct sem_undoundo;/*undo request on this array */
10)    unsigned short  sem_nsems;/*number of sem in array*/
11) };
12) /* the struct sem is defined as follows */
13) struct sem
14) {
15)    shortsempid;/* pid of last operation */
16)    ushortsemval;/* current val */
```

```
17)    ushortsemncnt /* num of process waiting semval > 0*/
18)    ushortsemzcnt;/* num of process waiting semval = 0 */
19) }
```

Semaphore Support Functions

The Linux kernel provides the following system calls for application programs to create and manage the semaphore data structures:

semget() Create a new semaphore, or access an existing one.

semop() Decrement, increment, or zero out the semaphore counts.

semctl() Allows control operation on the semaphore, such as getting the value of the semaphore, etc.

semget

```
Create a new semaphore or access an existing one
```

Prototype

```
int semget( key_t key, int nsems, int semflg);
```

Parameters

```
    key_t key:
The kernel tries to match this with exiting key values, the open or
access operation depends upon the flag semflg
    int nsems:
            The number of semaphores to be created in the set
    semflg:
            combination of the following types
            IPC_CREAT
                    Create the semaphore if does not exist already else
                    return the existing semaphore identifier
            IPC_EXCL
                    When used with IPC_CREAT, fails if semaphore already
                    exist
```

Returns

```
shared memory segment identifier
or –1 on error
    errno
```

```
= EACCESS (Permission denied)
= EEXIST (Segment exist, cannot create)
= EIDRM (segment reserved)
= ENOENT (Segment does not exist)
= ENOMEM (Not enough memory)
= ENOSPC (Maximum set limit exceeds)
```

semop

Increment or decrement the semaphore count. When the semaphore count reaches 0, the semaphore is locked and all processes are blocked.

If we take a simple case and initialize a semaphore with a count of 1, the decrement operation is like putting a lock on the semaphore, since the sem count is now 0. All subsequent decrement operations will be blocked and the calling process will be put to sleep until an increment operation is performed. See the man pages for details of the semop operation.

Prototype

```
int semop( int semid, struct sembuf * sops, unsigned nsops);
```

Parameters

```
    int semid:
The id returned from the semget
    Struct sembuf * sops:
Pointer to an array of operations to be performed on the semaphore set.
Struct sembuf
{
    ushort sem_num;         /* semaphore index in array */
    short semop;            /* semaphore operation pos, neg, zero */
    short sem_flg;          /* operation flags */
}
```

sem_num is the index of the semaphore in the semaphore set in which you are interested. If you have created only one semaphore, this value should be 0.

There are three possible scenarios for semop values:

- sem_op is a positive integer:
  The operation adds this value to semval. The operation always goes through, so no process sleeping can occur.

- sem_op is zero:
  If semval is zero, the operation goes through; otherwise, we have the following conditions based on the sem_flg:
    IPC_NOWAIT flag:
    The system call fails, errno set to EAGAIN.
    SEM_UNDO flag:
    semzcnt is incremented by 1, and the process sleeps.
- sem_op is less than zero:
  If semval is greater than or equal to the absolute value of sem_op, the absolute value of sem_op is subtracted by semval, the operation goes through, and no sleeping occurs. Otherwise, the process sleeps until semval is greater than or equal to the absolute value of sem_op.

The following is a simple example of sem_buf:

locking resources:

```
struct sembuf sem_lock = {0, -1, SEM_UNDO};
```

unlocking resources:

```
struct sembuf sem_unlock = {0, 1, SEM_UNDO};
```

Returns

If successful, the system call returns 0; otherwise, it returns –1, with errno indicating the error.

For a failing return, errno will be set to one of the following values:

E2BIG The argument nsops is greater than SEMOPM, the maximum number of operations allowed per system call.

EACCES The calling process has no access permissions on the semaphore set as required by one of the specified operations.

EAGAIN An operation could not go through, and IPC_NOWAIT was asserted in its sem_flg.

EFAULT The address pointed to by sops is not accessible.

EFBIG For some operation, the value of sem_num is less than 0, or greater than or equal to the number of semaphores in the set.

EIDRM The semaphore set was removed.

EINTR Sleeping on a wait queue, the process received a signal that had to be caught.

EINVAL The semaphore set does not exist, ID is less than 0, or nsops has a nonpositive value.

ENOMEM The sem_flg of some operation asserted SEM_UNDO, and the system does not have enough memory to allocate the undo structure.

ERANGE For some operation, semop+semval is greater than SEMVMX, the implementation-dependent maximum value for semval.

semctl

Allows control of the data structure, such as initializing the semaphore values, changing permissions, deleting the semaphore, and so forth.

Prototype

```
int semctl( int semid, int sem_num, int command, …);
```

Parameters

```
    int semid:
The id returned from the semget
    int sem_num:
            semaphore index in array
    int cmd:
            Two common commands are
            SETVAL
                    Initializing a semaphore to a known value
            IPC_RMID
                    Remove the semaphore from the system when it is no
                    longer needed.
```

Returns

The system call returns a nonnegative value depending on cmd as follows:

GETNCNT returns the value of semncnt.

GETPID returns the value of sempid.

GETVAL returns the value of semval.

GETZCNT returns the value of semzcnt.

On fail, the system call returns –1 with errno indicating the error.
For a failing return, errno will be set to one of the following values:

EACCES The calling process has no access permissions to execute cmd.

EFAULT The address pointed to by arg.buf or arg.array is not accessible.

EIDRM The semaphore set was removed.

EINVAL Invalid value for cmd or semid.

EPERM The argument cmd has value IPC_SET or IPC_RMID, but the calling process' effective user ID has insufficient privileges to execute the command.

ERANGE The argument cmd has value SETALL or SETVAL, and the value to which semval has to be set (for some semaphore of the set) is less than 0 or greater than the implementation value SEMVMX.

Using Semaphore in a Multiprogramming Environment

ON THE CD

We discussed the mutual exclusion in shared memory using semaphore, and now we will implement two test programs to demonstrate the synchronization of events using semaphores. The programs in Listings 4.7 and 4.8 are similar in execution (the source code is provided in the companion CD directory /book/chapter4/sem_x.c and /book/chapter4/sem_y.c). After creating a semaphore and executing a loop 10 times, the programs terminate. Each tries to acquire the semaphore and is being held until it is released from the other program.

- Task A gets a lock on the semaphore.
- Task A prints the message "Task A controls resources."
- Task A unlocks the semaphore.
- Task B gets a lock on the semaphore.
- Task B prints the message "Task B controls resources."
- Task B unlocks the semaphore.

LISTING 4.7 File /book/chapter4/shmem/sem_x.c; a program using semaphore for synchronization.

```
1) #include <stdio.h>
2) #include <unistd.h>
3) #include <stdlib.h>
4) #include <errno.h>
5) #include <string.h>
6) #include <time.h>
7) #include <sys/types.h>
8) #include <sys/ipc.h>
9) #include <sys/sem.h>
10) struct sembuf sem_lock = {0, -1, SEM_UNDO};
11) struct sembuf sem_unlock = {0, 1, SEM_UNDO};
12) static void end(const char *err)
13) {
```

```
14)    if (errno != 0)
15)    {
16)       fputs(strerror(errno),stderr);
17)       fputs(": " , stderr);
18)    }
19)    fputs(err,stderr);
20)    fputc('\n', stderr);
21) }
22) /*********************************************/
23) /*********************************************/
24) /*********************************************/
25) /*********************************************/
26) /*********************************************/
27) int main(int argc, char ** argv)
28) {
29)    int i, sem_id;
30)    int pause_time;
31)    srand (( unsigned int) getpid());
32)    sem_id = semget (( key_t) 1234, 1, 0666 | IPC_CREAT);
33)    /* initialize semaphore to a known unlock value 1 */
34)    if (semctl(sem_id, 0, SETVAL, 1) == -1)
35)       end("Failed to create semaphore ");
36)    for (i = 0; i < 10; i++)
37)    {
38)       if (semop(sem_id, &sem_lock, 1) == -1)
39)       end("Lock failed ");
40)       printf("Task A controls resources\n"); fflush(stdout);
41)       pause_time = (rand() % 3);
42)       sleep(pause_time);
43)       if (semop(sem_id, &sem_unlock, 1) == -1)
44)          end("Lock failed ");
45)    }
46)    sleep(10);
47)    /* Delete semaphore */
48)    if (semctl(sem_id, 0, IPC_RMID, 0) == -1)
49)       end("Failed to delete semaphore ");
50)    exit(0);
51) }
```

LISTING 4.8 File /book/chapter4/shmem/sem_y.c; a program using semaphore for synchronization.

```
1)  #include <stdio.h>
2)  #include <unistd.h>
```

```
3) #include <stdlib.h>
4) #include <errno.h>
5) #include <string.h>
6) #include <time.h>
7) #include <sys/types.h>
8) #include <sys/ipc.h>
9) #include <sys/sem.h>
10) struct sembuf sem_lock = {0, -1, SEM_UNDO};
11) struct sembuf sem_unlock = {0, 1, SEM_UNDO};
12) static void end(const char *err)
13) {
14)    if (errno != 0)
15)    {
16)       fputs(strerror(errno),stderr);
17)       fputs(": ", stderr);
18)    }
19)    fputs(err,stderr);
20)    fputc('\n', stderr);
21) }
22) /**********************************************/
23) /**********************************************/
24) /**********************************************/
25) /**********************************************/
26) /**********************************************/
27) int main(int argc, char ** argv)
28) {
29)    int i, sem_id;
30)    int pause_time;
31)    srand (( unsigned int) getpid());
32)    sem_id = semget (( key_t) 1234, 1, 0666 | IPC_CREAT);
33)    /* initialize semaphore to a known unlock value 1 */
34)    if (semctl(sem_id, 0, SETVAL, 1) == -1)
35)       end("Failed to create semaphore ");
36)    for (i = 0; i < 10; i++)
37)    {
38)       if (semop(sem_id, &sem_lock, 1) == -1)
39)       end("Lock failed ");
40)       printf("Task B controls resources\n"); fflush(stdout);
41)       pause_time = (rand() % 3);
42)       sleep(pause_time);
43)       if (semop(sem_id, &sem_unlock, 1) == -1)
44)          end("Lock failed ");
45)    }
46)    sleep(10);
```

```
47)    /* Delete semaphore */
48)    if (semctl(sem_id, 0, IPC_RMID, 0) == -1)
49)       end("Failed to delete semaphore ");
50)    exit(0);
51) }
```

ON THE CD

Use the Makefile provided in Listing 4.9 (CD directory /book/chapter4/semaphore/Makefile) to compile the sem_x.c and sem_y.c programs, and execute the two programs in separate shells using the following shell command:

```
# ./x.o
```

Open another shell and execute sem_y.o

```
#./y.o
```

Each program will display a message once semaphore is being obtained.

LISTING 4.9 File /book/chapter4/semaphore/Makefile; a Makefile to create the sem_x.o and sem_y.o executable.

```
1)  all: t
2)  CC = gcc
3)  CFLAGS = -g -02 -Wall -I/usr/include -I/usr/include/pgsql
4)  t: sem_x.o sem_y.o
5)     $(CC) -o x sem_x.o -lpq -lcrypt
6)     $(CC) -o y sem_y.o -lpq -lcrypt
7)  sem_x.o: sem_x.c
8)     $(CC) -o sem_x.o -c sem_x.c $(CFLAGS)
9)  sem_y.o: sem_y.c
10)    $(CC) -o sem_y.o -c sem_y.c $(CFLAGS)
```

MESSAGE QUEUES

Message queues provide the capability to send and receive a block of data from one process to another. However, they are unique in one aspect in that one process can be made to wait on messages from several sources. Each message in the message queue has a unique type associated with it, and processes can selectively receive a specific message while ignoring others.

Similar to the shmid_ds structure of the shared memory, the kernel maintains a special internal data structure msqid_ds for message queues also in the include file <sys/ipc.h>.

```
1) struct msqid_ds
2) {
3)     struct ipc_perm sem_perm;/*permission */
4)     struct msg * msg_first;/* first msg in the queue */
5)     struct msg * msg_last;/* last msg in the queue */
6)     time_tmsg_stime;/* last msg send time */
7)     time_tmsg_rtime;/* last msg rcv time */
8)     time_tmsg_ctime;/* last msg change time */
9)     struct wait_queue wwait;struct wait_queue rwait;
10)    ushort msg_cbytes;
11)    ushort msg_qnum;
12)    ushort msg_qbytes;/* max number of bytes on queue */
13)    ushort msg_lspid;/* pid of last send */
14)    ushort msg_lrpid;/* pid of last rcv */
15) }
```

Message queues are identified by an IPC identifier, while the message and its type are embedded in a structure called msgbuf, defined here:

```
#include <linux/msg.h>
/* message buffer for send and receive */
struct msgbuf
{
    long mtype;     /*type of message, a positive number */
    char mtext[1]; /* message text */
}
```

The struct msgbuf is only a template, and the member mtext may be replaced by user-defined values.

The kernel stores each message in the form of a linked list as described here:

```
struct msg
{
    struct msg * msg_next;      /* next msg in the queue */
    long msg_type;              /*type of message, a positive number */
    char *msg_text;             /* message text */
    short msg_ts;               /* msg text size <= 4096 */
}
```

Sending a Message

```
int msgsnd( int msqid, struct msgbuf *msgp, int msgsz, int msgflg)
Return:

    0 on success;
    -1 on Error, errno is one of the following
          EACCESS (permission denied)
          EAGAIN (IPC_NOWAIT Asserted)
          EIDRM (Semaphore set was removed)
          EINTR (signal received while sleeping )
          EINVAL (sem_id is invalid )
          ENOMEM (not enough memory)
```

Receiving a Message

```
int msgrcv( int msqid, struct msgbuf *msgp, int msgsz, long mtype, int
msgflg)
Return:
    0 on success;
    -1 on Error, errno is one of the following
          E2BIG (nsops greater then max number of operations allowed)
          EACCESS (permission denied)
          EAGAIN (IPC_NOWAIT Asserted)
          EIDRM (message set was removed)
          EINTR (signal received while sleeping )
          EINVAL (msqid is invalid )
          ENOMSG (no msg exist)
```

Using Message Queues in a Multiprogramming Environment

ON THE CD

Message queues provide a convenient method for synchronizing events among tasks just as semaphores, but the message queues have the advantage of passing parameters also. The fact that a task can be forced to wait on a specific message can be used in protecting shared resources. Listings 4.10 and 4.11 are two programs (the source files can be downloaded from the companion CD directory /book/chapter4/msgque.c for Listing 4.9, and /book/chapter4/test/test.c for Listing 4.10). The programs work in conjunction and pass messages to each other, using a message queue created by the msgget (line 21). Each program sends a message to the other (line 29) and then waits on a message from others (line 35).

LISTING 4.10 File /book/chapter4/msgqueues/msgque.c; a program using message queues for communication.

```
1) #include <unistd.h>
2) #include <stdlib.h>
3) #include <stdio.h>
4) #include <string.h>
5) #include <sys/types.h>
6) #include <sys/ipc.h>
7) #include <sys/msg.h>
8) #define MSG_FROM_A 1
9) #define MSG_FROM_B 2
10) #define MESSAGE_KEY 9988
11) #define MSG_SIZE 80
12) struct msgbuf
13) {
14)    int mtype;
15)    char bfr[MSG_SIZE];
16) };
17) struct msgbuf qbuf;
18) int msgid;
19) int main()
20) {
21)    msgid = msgget (MESSAGE_KEY, 0666 | IPC_CREAT);
22)    if (msgid == -1)
23)    {
24)       fprintf(stderr, "msg queue failed \n");
25)       exit (EXIT_FAILURE);
26)    }
27)    qbuf.mtype = MSG_FROM_A;
28)    strcpy(qbuf.bfr,"Hello Task B:\n");
29)    if (msgsnd(msgid, &qbuf, MSG_SIZE, 0) == -1)
30)    {
31)       fprintf(stderr, "msg queue failed \n");
32)       exit (EXIT_FAILURE);
33)    }
34)    printf ("Waiting for a message from Task B\n");
35)    if (msgrcv(msgid, &qbuf, MSG_SIZE, MSG_FROM_B, 0) == -1)
36)    {
37)       fprintf(stderr, "msg queue failed \n");
38)       exit (EXIT_FAILURE);
39)    }
40)    printf("Msg from B: %s\n",qbuf.bfr);
41)    printf("Task A deleting the msg queue\n");
```

```
42)    if (msgctl(msgid, IPC_RMID, 0) == -1)
43)    {
44)       fprintf(stderr, "Could not delete msg queue \n");
45)       exit (EXIT_FAILURE);
46)    }
47)    exit (EXIT_SUCCESS);
48) }
```

LISTING 4.11 File /book/chapter4/msgqueues/test/test.c; a companion program for Listing 4.10, testing the message passing scheme.

```
1)     #include <unistd.h>
2)     #include <stdlib.h>
3)     #include <stdio.h>
4)     #include <string.h>
5)     #include <sys/types.h>
6)     #include <sys/ipc.h>
7)     #include <sys/msg.h>
8)     #define MSG_FROM_A 1
9)     #define MSG_FROM_B 2
10)    #define MESSAGE_KEY 9988
11)    #define MSG_SIZE 80
12)    struct msgbuf
13)    {
14)       int mtype;
15)       char bfr[MSG_SIZE];
16)    };
17)    struct msqid_ds queue_ds;
18)    struct msgbuf qbuf;
19)    int msgid;
20)    int main()
21)    {
22)       msgid = msgget (MESSAGE_KEY, 0666 | IPC_CREAT);
23)       if (msgid == -1)
24)       {
25)          fprintf(stderr, "msg queue failed \n");
26)          exit (EXIT_FAILURE);
27)       }
28)       msgctl(msgid, IPC_STAT, &queue_ds);
29)       printf("Msg queue permission setting %x\n",queue_ds.msg_perm.mode);
30)       printf ("Waiting for a message from Task A\n");
31)       if (msgrcv(msgid, &qbuf, MSG_SIZE, MSG_FROM_A, 0) == -1)
32)       {
```

```
33)         fprintf(stderr, "msg queue failed \n");
34)         exit (EXIT_FAILURE);
35)     }
36)     printf("Msg from A: %s\n",qbuf.bfr);
37)     qbuf.mtype = MSG_FROM_B;
38)     strcpy(qbuf.bfr,"Hello Task A:\n");
39)     if (msgsnd(msgid, &qbuf, MSG_SIZE, 0) == -1)
40)     {
41)         fprintf(stderr, "msg queue failed \n");
42)         exit (EXIT_FAILURE);
43)     }
44)     exit (EXIT_SUCCESS);
45) }
```

Compile the msgque.c program using the Makefile in Listing 4.12, and compile the test.c program using the Makefile in Listing 4.13.

LISTING 4.12 File /book/chapter4/msgques/Makefile; a Makefile to create the msgque.c executable.

```
1) all: msgque
2) CC = gcc
3) INCLUDEDIR = /usr/include
4) CFLAGS = -g -02 -Wall -I$(INCLUDEDIR)
5) msgque: msgque.o
6)    $(CC) -o t msgque.o
7) msgque.o: msgque.c
8)    $(CC) -c msgque.c $(CFLAGS)
```

LISTING 4.13 File /book/chapter4/msgques/test/Makefile; a Makefile to create the test.o executable.

```
1) all: t
2) CC = gcc
3) INCLUDEDIR = /usr/include
4) CFLAGS = -g -02 -Wall -I$(INCLUDEDIR)
5) t: test.o
6) $(CC) -o t test.o
7) test.o: test.c
8) $(CC) -c test.c $(CFLAGS)
```

Execute the two programs msgque.o and test.o in their respective shells and watch the cross message displayed by each other using the following shell command:

```
# ./t
```

Open another shell in the test directory and execute the test.o.

```
#./t
```

Each program will display messages received from the other program.

PIPES

Pipes are mainly used to connect the standard output of one process to the standard input of another. It is basically a one-way communication. The shell interprets the "|" key as the piping command to redirect the output of one process to the input of another one. For example, the following shell command will display only those filenames that contain the string Mak:

```
#  find . | grepMak
```

In a software program, pipes are created and closed as any other file; the following function declaration is included in the stdio.h header file:

FILE * popen(char * command, char * type);

int pclose(FILE * stream);

The standard library function executes the popen() command by essentially forking a child process: spawning a shell, and execute the argument passed to the function popen(). The direction of the data flow is determined by the second argument.

The pclose() function performs a wait4() on the process forked by popen() before closing the stream.

ON THE CD

Listing 4.14 is essentially the same as the shell command of finding all the filenames that contain the string 'Mak.' The source and the Makefile for Listing 4.14 can be copied from the CD directory /mnt/cdrom/book/chapter4/pipes/Makefile.

Compile and execute the pipes.c (Listing 4.14) program using the following shell command

```
# mkdir /pipes
# cp /mnt/cdrom/book/chapter4/pipes/* /pipes/.
# cd /pipes
```

```
# make
# ./t
```

The program will display file names with the string "Mak" starting from the root directory.

LISTING 4.14 File /book/chapter4/pipes/pipe.c; a sample program showing the use of pipe function

```
1) #include <unistd.h>
2) #include <stdlib.h>
3) #include <stdio.h>
4) #include <string.h>
5) #include <sys/types.h>
6) #include <sys/ipc.h>
7) #include <sys/msg.h>
8) int main()
9) {
10)    char buf[80];
11)    FILE * pin, * pout;
12)    /* open a pipe for read. Exit if fail */
13)    if  ((pin = popen("cd / ; find . ", "r"))  == NULL)
14)    {
15)        printf("pipe open error \n");
16)        exit(1);
17)    }
18)    /* open a pipe for write. Exit if fail */
19)    if  ((pout = popen("grep Mak", "w"))  == NULL)
20)    {
21)        printf("pipe open error \n");
22)        exit(1);
23)    }
24)    while (fgets(buf, 80, pin))
25)       fputs(buf,pout);
26)    exit(0);
27) }
```

FORK

ON THE CD

The fork function essentially duplicates a process image; one remains the parent, and the other is the child. The parent and child processes become two separate instances of the same program that do not share the same memory space once the fork is called. The program creates a perception that a variable is present inside the scope of a function, but it is actually not. To enhance your understanding of the fork process, the program in Listing 4.15 (you can copy the file from the companion CD directory /book/chapter4/fork/fork.c) creates a child process (line 24) that, in turn, creates another child process (line 29). Thus, there are three instances of the same program all executing in a different memory space. Compile the program using the Makefile shown in Listing 4.16. The Makefile can be copied from the companion CD directory /book/chapter4/fork/Makefile.

LISTING 4.15 File /book/chapter4/fork/fork.c; a program showing fork function creating three instances of the same program.

```
 1) #include <stdio.h>
 2) #include <unistd.h>
 3) #include <stdlib.h>
 4) #include <errno.h>
 5) #include <string.h>
 6) #include <time.h>
 7) #include <sys/types.h>
 8) #include <sys/socket.h>
 9) #include <arpa/inet.h>
10) #include <netdb.h>
11) static void end(const char *err)
12) {
13)    if (errno != 0)
14)    {
15)       fputs(strerror(errno),stderr);
16)       fputs(": ", stderr);
17)    }
18)    fputs(err,stderr);
19)    fputc('\n', stderr);
20) }
21) int main(int argc, char ** argv)
22) {
23)    pid_t cpid=0, ppid=0;
24)    ppid = fork();
25)    switch(ppid)
26)    {
27)    case 0:
28)       printf("This is child 1st child pid %d\n",ppid);
29)       cpid = fork();
30)       printf("This is child 2nd child pid:%d\n",cpid);
```

```
31)        break;
32)     case -1:
33)        end("Failed fork");
34)     default:
35)        printf("This is parent pid:%d\n",ppid);
36)     break;
37)     }
38)     sleep(3);
39)     printf("Finish ppid:%d, cpid:%d\n",ppid,cpid);
40)     return 0;
41) }
```

LISTING 4.16 File /book/chapter4/fork/Makefile; the Makefile to create fork.o program.

```
1)  all: t
2)  CC = gcc
3)  CFLAGS = -g -02 -Wall -I/usr/include -I/usr/include/pgsql
4)  t: fork.o
5)     $(CC) -o t fork.o -lpq -lcrypt
6)  fork.o: fork.c
7)     $(CC) -o fork.o -c fork.c $(CFLAGS)
```

Execute the fork program using the following shell program:

```
# ./t
```

You will see the following messages from the three instances:

```
This is parent pid 551
This is child 1st child pid 0
This is child 2nd child pid 552
```

SOCKETS

Among all the methods available for interprocess communication—pipes, sockets, signals, semaphores, and queues—sockets are by far the most versatile. They break all the barriers of distance. Two processes can be running on the same local CPU, or be miles apart running on different machines. Sockets provide a seamless end-to-end communication method. Like all other systems in Linux, the sockets are also treated as file descriptors. Once sockets are created, we can perform the same read and write operations as we do with the disk file systems. The only difference is that the disk files are attached to the nodes on the file system, and the sockets are attached to the IP addresses of their respective machines.

Establishing a communication link using sockets is much like making telephone calls. You want to establish a link: you dial the number of the other party, the other party hears a ring, the party picks up the telephone, and you have an end-to-end communication. Using this analogy, think of the caller as the *client*, the listener as the *server*, and the telephone numbers as the *IP addresses*. A *port* on the IP address is like an extension number in a local exchange. Unless there is a server (the listener) on the other side, waiting for a telephone call, the client (the caller) cannot establish communication. Thus, a socket server is different from the socket client, as the server, once created, should simply *listen* for the call, while a client should *connect* to the server after being created.

The roles of the client and the server sockets might be different, but the method of creation is the same for both. Sockets are created using the operating system function socket(), but they must be bound to a file node using the function bind() before any communication is performed. Once a socket is bound, the same read, write, open, and close file operations that you have on regular files are available. There are subtle differences, however, and Figure 4.3 explains the different stages each socket goes through.

Socket Creation

The Linux operating system provides the following function call for creating sockets:

```
int socket (int domain, int type, int protocol);
```

We discuss the three arguments—the domain, the type, and the protocol—in the following sections.

Domains

The sockets do not have to reside on different machines; you can create both (client and server) locally on the same machine. Sockets definitely need addresses for remote connections, while local connections do not. Sometimes, the client socket on the remote connection does not need an address, but the server does. Also remember that TCP/IP is not the only protocol for socket communication; there are many more as listed in Table 4.1. All this is considered *domain* type. For local sockets, it is the AF_LOCAL domain; for remote TCP/IP, it is the AF_INET domain; for a Novell network, it is the AF_IPX domain, and so forth. (The AF stands for address family, but the prefix PF is also being used, such as PF_LOCAL, PF_INET, and so

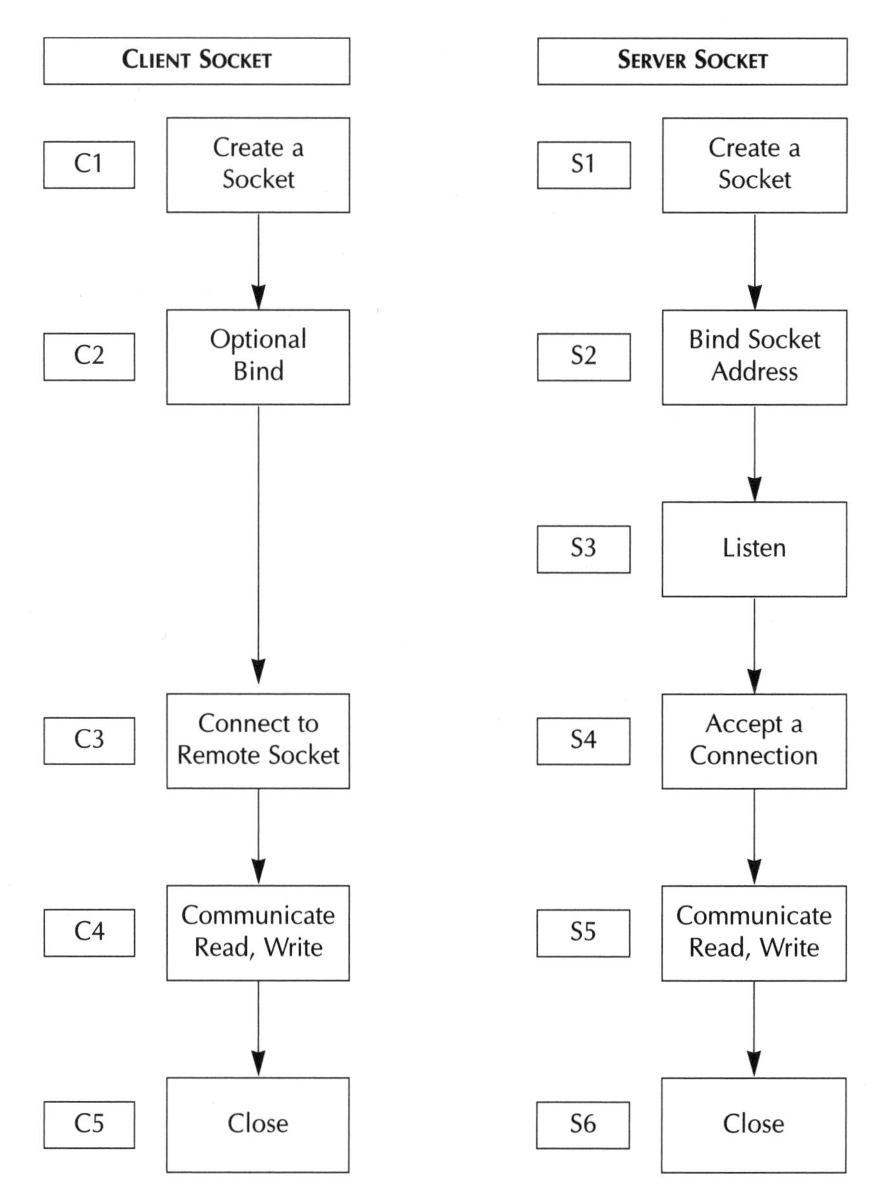

FIGURE 4.3 The different stages in the life cycle of a client and a server socket.

forth.) The domain information is important to the operating system when the request to create a socket is presented. You can also leave the request blank and let the operating system pick the right one for you.

TABLE 4.1 List of Socket Domains

PF_UNIX,PF_LOCAL	Local communication	unix(7)
PF_INET	IPv4 Internet protocols	ip(7)
PF_INET6	IPv6 Internet protocols	
PF_IPX	IPX - Novell protocols	
PF_NETLINK	Kernel user interface device	netlink(7)
PF_X25	ITU-T X.25 / ISO-8208 protocol	x25(7)
PF_AX25	Amateur radio AX.25 protocol	
PF_ATMPVC	Access to raw ATM PVCs	
PF_APPLETALK	AppleTalk	ddp(7)
PF_PACKET	Low level packet interface	packet(7)

The *domain* argument defines the protocol family to use; the two most common are PF_LOCAL and PF_INET, and other less common protocols are listed in Table 4.1.

Socket Communication Type

Sockets are not restricted to a specific type. You can specify a reliable error-free connection, a not very reliable broadcast connection, a one-way, two-way, sequenced, out-of-band connection, and so forth. The *type* argument specifies the communication semantics. Currently defined types are:

SOCK_STREAM Provides sequenced, reliable, two-way, connection-based byte streams. An out-of-band data transmission mechanism may be supported.

SOCK_DGRAM Supports datagrams (connectionless, unreliable messages of a fixed maximum length).

SOCK_SEQPACKET Provides a sequenced, reliable, two-way connection-based data transmission path for datagrams of fixed maximum length; a consumer is required to read an entire packet with each read system call.

SOCK_RAW Provides raw network protocol access.

SOCK_RDM Provides a reliable datagram layer that does not guarantee ordering.

SOCK_PACKET Obsolete and should not be used in new programs; see packet(7). Some socket types may not be implemented by all protocol families; for example, SOCK_SEQPACKET is not implemented for AF_INET.

Socket Protocol

When you specify the socket domain at the time of socket creation, some domains inherently specify the type of protocol to use; for example, AF_INET automatically selects TCP/IP. Other domains might require you to specify the protocol type. The *protocol* argument specifies a particular protocol to use with the socket. The default value 0 means the operating system will pick the matching protocol. Normally, only a single protocol exists to support a particular socket type within a given protocol family. However, it is possible that many protocols might exist, in which case a particular protocol must be specified. The protocol number to use is specific to the "communication domain." You can find a mapping of protocols in the /etc/protocols file.

Binding a Socket to an Address

Once a socket is created, it should be bound to an address; especially, if it is a server socket. That address is dependent on to whom and where the connection is being made. If your sockets are supposed to communicate locally on the same machine, the address is a file path. However, if the sockets are on two separate machines, the address is the network address. Linux provides the following function calls for binding a socket to an address:

```
int  bind  (int  socket,  struct  (sockaddr *)  &sock_address,  int len_inet);
```

The first argument *socket* is the socket ID returned from the function socket(). The second argument is the pointer to a structure sockaddr. The structure itself is not that important; you could pass along anything, but make sure that the first two bytes indicate the address family you intend to use. Whatever address structure you choose, the first two bytes determine how the rest of the bytes are interpreted. Think of the first two bytes as the type specifying how the rest of the structure will appear. The structure provides only guidelines for forming an address; other address structures are shown here:

Generic address:

```
struct sockaddr
{
    sa_family_t    sa_family;     /* a 2 byte short Address Family */
    char           sa_data[14];   /* Address data */
};
```

Local address:

```
The structure name for AF_LOCAL or AF_UNIX address is struct
sockaddr_un
{
    sa_family_t    sun_family; /* Address family */
    char sun_path[108];
};
```

Internet socket address:

```
The structure name for AF_INET is struct sockaddr_in
{
    sa_family_t    sin_family; /* Address family */
    uint16_t       sin_port;
    struct_in_addr sin_addr;      /* Internet Address */
    unsigned char sin_zero[8];   /* pad bytes */
};
```

Network Byte Ordering

A byte is a byte; however, when it comes to words, there are two types of words in this world, a Motorola type of word and an Intel type of word. The difference is the way a word (two bytes) is written into the computer's memory. This can be explained as the CPU architecture dependency.

Motorola

Big Endian

Hexadecimal value 0x1234 in a 16-bit integer is 0x12,0x34.

Intel

Little Endian

Hexadecimal value 0x1234 in a 16-bit integer is 0x34,0x12.

The orientation makes a difference, and you must follow the Motorola type (Big Endian) if your socket has to communicate over the Internet. The truth is, the Big Endian arithmetic offers a slightly advantage over the Little Endian. If the memory is read one byte at a time, reading the most significant byte into the memory provides an early start in some long arithmetic operations. Conversion functions provided by the operating system of the platform are available for performing correct orientations. The following is a list of conversion functions:

```
#include <netinet/in.h>
unsigned long htonl (unsigned long hostlong)
unsigned short htons (unsigned short hostshort)
unsigned long ntonl (unsigned long netlong)
unsigned long ntons (unsigned short netshort)
```

IP Address Convention

An IP address is a 32-bit number specifying the host and network ID combined. A network is comprised of a group of hosts. For example, if a network number is 0x10, then 0x10000001 through 0x10FFFFFF are all host numbers. A netmask is available to isolate the host portion from the IP address. Conventionally, the 32-bit (4-byte) number is written as a group of four dotted decimal numbers, such as 200.200.200.200. The entire IP addressing system needs a big overhaul, as IP addresses on the Internet are becoming scarce, because of the way the network portion of the IP addressing is carved out. Theoretically, the 32-bit IP addressing scheme can provide connections to approximately 4 billion computers—apparently a very large number. However, because of the way in which the network portions are doled out, the first 126 organizations receive 13 million host numbers each, for a total of approximately 2 billion numbers (category A numbers). The next 65,000 organizations receive 65,000 each, for a total of approximately 1 billion numbers (category B numbers). The remaining 13 million organizations (category C) receive 256 host numbers each—there goes the complete quota.

There is an alternate scheme of IPv6 that allows a 128-bit numbering scheme, but it has yet to become popular. The following is the breakdown of the 32-bit IPv4 IP addressing scheme. The XX is the host portion of the IP address. In this portion, some numbers are reserved for special purposes: 0.XX.XX.XX is reserved for the default route; 127.XX.XX.XX is reserved for local access; 16.XX.XX.XX and 192.168.XX.XX are reserved for private networks; and, if the first three bits are 111, such as FF.FF.FF.FF, then it is a multicast address. Multicast addresses are used to address groups of computers simultaneously. In all networks, hosts 0 and 255 are reserved. An IP address with all host bits set to 0 identifies the network itself. An IP address with all bits set to 1 is a broadcast address.

Category A	01XXXXXX Ö 7EXXXXXX	(last bit is 0)
Category B	8000XXXX Ö BFFFXXXX	(last two bits are 10)
Category C	C00000XX Ö DEFFFFXX	(last three bits are 110)
Category D	EXXXXXXX Ö FXXXXXX	(last three bits are 111)

Nameless Sockets

Socket addresses are like telephone numbers, and do not always have to have an address; for example, if there is a direct line between two sockets. Socketpair() is an example of creating two sockets without addresses. Sometimes, one of the two sockets in communication will have no address. A local socket that is placing the call can be anonymous.

Server Socket Listens for Clients

If it is a server socket, it should create a connection queue and should be waiting for clients to make connections. The function listen() provides this facility. The listen call applies only to sockets of type SOCK_STREAM or SOCK_SEQPACKET. The function prototype is shown here:

```
#include <sys/socket.h>
int listen(int s, int backlog);
```

Description The backlog parameter defines the maximum length the queue of pending connections. The argument s is the socket number.

Return Value_On success, 0 is returned; on error, –1 is returned; and errno is set appropriately.

Errors

EBADF The argument s is not a valid descriptor.

ENOTSOCK The argument s is not a socket.

EOPNOTSUPP The socket is not of a type that supports the listen operation.

Client Connecting to the Server

If it is a client socket, it should be connected to the desired server. The function connect() provides this facility. The function prototype is shown here:

```
#include <sys/types.h>
#include <sys/socket.h>
int  connect(int sockfd, const struct sockaddr *serv_addr,
socklen_t addrlen);
```

Description The file descriptor sockfd must refer to a socket. If the socket is of type SOCK_DGRAM, the serv_addr address is the address to which datagrams are sent by default, and the only address from which datagrams are received. If the socket is of type SOCK_STREAM or SOCK_SEQPACKET, this call attempts to make a connection to another socket. The other socket is specified byserv_addr, which is an address (of length addrlen) in the communication space of the socket. Each communication space interprets the serv_addr parameter in its own way.

Server Accepting the Client Socket

Once a successful link is started, the server socket must accept the client specifically. This step allows a server to respond to several clients. The following is the function prototype:

```
#include <sys/types.h>
#include <sys/socket.h>
int   accept(int    s,  struct  sockaddr *addr,  socklen_t *addrlen);
```

Description The accept function is used with connection-based socket types (SOCK_STREAM, SOCK_SEQPACKET, and SOCK_RDM). It extracts the first connection request on the queue of pending connections, creates a new connected socket with mostly the same properties as s, and allocates a new file descriptor for the socket, which is returned. The newly created socket is no longer in the listening state. The original socket s is unaffected by this call. The argument s is a socket that has been created with socket(2), bound to a local address with bind(2), and is listening for connections after a listen(2). The argument addr is a pointer to a sockaddr structure. This structure is filled in with the address of the connecting entity, as known to the communication layer. The exact format of the address passed in the addr parameter is determined by the socket's family (see socket(2) and the respective protocol man pages). The addrlen argument is a value-result parameter: it should initially contain the size of the structure pointed to by addr; on return, it will contain the actual length (in bytes) of the address returned. When addr is NULL, nothing is filled in. If no pending connections are present in the queue, and the socket is not marked as nonblocking, accept blocks the caller until a connection is present. If the socket is marked nonblocking and no pending connections are present in the queue, accept returns EAGAIN.

Return Value The call returns –1 on error. If it succeeds, it returns a non-negative integer that is a descriptor for the accepted socket.

File Read/Write Operations

The descriptor returned by the function call accept() is being used to perform the normal file read and write operations on the server sockets. The client socket read and write is being performed using the descriptor that was returned by the socket function itself. Sockets are very similar to file descriptors. Once a socket is open, the same read, write, and close function calls can be applied. A file descriptor is used to refer to an opened file or a network socket. There are some differences, however:

- There is no lseek function on open socket files.
- Sockets can have addresses associated with them.
- Sockets have different option capabilities that can be queried and set using ioctl.
- Socket must be in the correct state to perform input and output.

Closing Sockets

Closing the socket file descriptors is a little different as well. Once you are finished with the writing and you want to close the file, you should wait for an acknowledgment from the other end; for this, the function "shutdown" is being used as follows:

```
#include <sys/socket.h>
int shutdown(in s, int how);
The s argument is the socket descriptor and the how argument is
SHUT_RD, SHUT_WR, SHUT_RDWR
```

Shutting down the writing end (SHUT_WR) sends an end-of-file indication to the remote connection.

Keep reading from the other end until you get an end-of-file, and then proceed with close. Shutting down the reading end (SHUT_RD) will start ignoring the received data.

Shutdown should only be used for connected sockets.

A Simple Web Server

ON THE CD

To demonstrate the socket mechanism, we will create a very simple Web server (Listing 4.17). The source code is available on the companion CD directory /book/chapter4/web/socket/web.c. Copy the source and the Makefile to a local directory /web. The server will generate HTML pages upon request from a Web browser such as Netscape Navigator or Internet Explorer connected to port 9999. The server will be used in the design of a data acquisition system later in Chapter 9, "Embedded System Design Projects." The following is an explanation of the software flow:

- Lines 16 through 22 are the routine to display an error and exit.
- Line 36 creates the socket.
- Line 49 binds the socket.

- Line 55 setsockopt allows the address to be reused.
- Line 62 makes it a listening socket (our program runs as a server).
- Lines 71 through 108 are a forever loop construct.
- Line 77 calls accept routine and returns only when a request is on the socket port 9999.
- Line 95 through 105 is a simple Web page that is being returned in response to the request.

LISTING 4.17 File /book/chapter4/socket/web.c; a simple Web server.

```
1)   /* web.c :
2)   A simple web server
3)   */
4)   #include <stdio.h>
5)   #include <unistd.h>
6)   #include <stdlib.h>
7)   #include <errno.h>
8)   #include <string.h>
9)   #include <fcntl.h>
10)  #include <time.h>
11)  #include <sys/types.h>
12)  #include <sys/socket.h>
13)  #include <netinet/in.h>
14)  #include <sys/un.h>
15)  #include <sys/uio.h>
16)  void end(const char *errmsg)
17)  {
18)     if ( errno != 0 )
19)         fprintf(stderr,"%s: ",strerror(errno));
20)     fprintf(stderr,"%s\n",errmsg);
21)     exit(1);
22)  }
23)  int main()
24)  {
25)     int z;
26)     int s;                /* Web Server socket */
27)     int c;                    /* Client socket */
28)     int alen;               /* Address length */
29)     struct sockaddr_in a_web; /* Web Server */
30)     struct sockaddr_in a_cln;/* Client addr */
31)     int b = 1;        /* For SO_REUSEADDR */
32)     FILE *rx;                   /* Read Stream */
33)     FILE *tx;                  /* Write Stream */
```

```
34)     char getbuf[2048];          /* GET buffer */
35)     time_t td;        /* Current date & time */
36)     s = socket(PF_INET,SOCK_STREAM,0);
37)     if ( s == -1 )
38)       end("socket(2)");
39)  /*
40)  Web address on port 9999:
41)  */
42)     memset(&a_web,0,sizeof a_web);
43)     a_web.sin_family = AF_INET;
44)     a_web.sin_port = ntohs(9999);
45)     a_web.sin_addr.s_addr =  ntohl(INADDR_ANY);
46)  /*
47)  Bind the web server address-
48)  */
49)     z = bind(s, (struct sockaddr *)&a_web, sizeof a_web);
50)     if ( z == -1 )
51)        end("binding port 9999");
52)  /*
53)  Turn on SO_REUSEADDR :
54)  */
55)     z = setsockopt(s,SOL_SOCKET,
56)     SO_REUSEADDR,&b,sizeof b);
57)     if ( z == -1 )
58)        end("setsockopt(2)");
59)  /*
60)  Now make this a listening socket:
61)  */
62)     z = listen(s,10);
63)     if ( z == -1 )
64)        end("listen(2)");
65)  /*
66)  Peform a simple, web server loop
67)  accept one line of input text, and
68)  ignore it. Provide one simple
69)  HTML page back in response:
70)  */
71)     for (;;)
72)     {
73)     /*
74)        Wait for a connect from browser:
75)     */
76)        alen = sizeof a_cln;
77)        c = accept(s, (struct sockaddr *)&a_cln,
```

```
78)                 &alen);
79)           if ( c == -1 )
80)           {
81)              perror("accept(2)");
82)              continue;
83)           }
84)     /*
85)     Create streams
86)     */
87)           rx = fdopen(c,"r");
88)           tx = fdopen(dup(c),"w");
89)           fgets(getbuf,sizeof getbuf,rx);
90)     /*
91)     Now serve a simple HTML response.
92)     This includes the current date
93)     and time:
94)     */
95)           fputs("<HTML>\n"
96)           "<HEAD>\n"
97)           "<TITLE>Test Page </TITLE>\n"
98)           "</HEAD>\n"
99)           "<BODY>\n"
100)          "<H1>Welcome!</H1>\n",tx);
101)          time(&td);
102)          fprintf(tx,"<H2>Current time: %s</H2>\n",
103)                ctime(&td));
104)          fputs("</BODY>\n"
105)          "</HTML>\n",tx);
106)          fclose(tx);
107)          fclose(rx);
108)       }
109)       return 0;
110)   }
```

Testing the Web Server

Compile the program in Listing 4.17 using the following command line:

```
# cd web
# make
```

Execute the program with the following command line. Make sure you shut down the existing HTTP server.

```
./web
```

You can communicate with the server by opening Netscape Navigator and requesting the local host. The welcome page will appear.

```
netscape 127.0.0.1:9999
```

SUMMARY

In this chapter, we discussed the concept of multitasking and how it helps us to modularize a computer program into manageable tasks. We created example programs by spawning multiple tasks that form the basis of a modular embedded system design. We also discussed the different interprocess communication facilities provided by the Linux operating system, including shared memory, semaphores, message queues, pipes, fork, and sockets. We developed a small Web server to demonstrate the mechanism of providing communication among networked computers such as the ones you see through the Internet.

5 Perl Programming

In this chapter

- A Short Course on Perl
- Functions and Subroutines
- Local Variables
- Perl Statements
- Basic Blocks and Switch Statements
- Perl Modules
- PG.PM
- Built-in Functions in Perl

INTRODUCTION

The Practical Extraction and Report Language, or Perl, is ideal for designing interactive Web pages for Internet use; thus, our user interface needs Perl. A user interface for an embedded system is about monitoring the processed data in real time, and if you are using the Internet to view the data, it is only possible to do so with an interpretive language such as Perl. Browsers such as Netscape only understand pages in HTML syntax; these are merely static text files. The first one is usually stored as "index.html" in a computer named "www" in a fictitious domain "company.com." That is why you enter http://www.company.com on the browser's request line. When the request reaches the destination Web server, the file contents are simply transferred to the requested browser without interpretation. On the other hand, when a link is requested to a file that has a .pl or .cgi extension (see the ExecCGI configuration in /etc/httpd/conf/httpd.conf file), the Web server invokes the appropriate interpreter to interpret the contents of the file or the script as it is being called. This is where Perl comes in to play. The text generated by the printf statements within the script is the text that is transferred to the requesting browser. It is a scriptwriter's job to provide appropriate statements in the script that generate a proper HTML syntax to produce the desired Web page. You will see how Perl helps you in creating dynamic Web pages through script interpretation later in the chapter.

Credit goes to Larry Wall for designing Perl, a simple, yet powerful language and ideal for developing Web-based software. Without knowing much about the programming aspects of Perl, you can create HTML pages and common gateway interface (CGI) scripts that interface to the database—this is the beauty of object-oriented programming. There is an extensive library of Perl modules in CPAN (Comprehensive Perl Archive Network) for almost every aspect of programming. A user simply has to understand the interface, and the Perl module takes care of the rest of the programming detail. In this chapter, we will cover the Perl CGI and Pg modules, but for more in-depth information, please visit the official Web site at www.perl.com/pub/v/documentation/.

Most Linux distributions install Perl by default, but if your system lacks the interpreter, please visit the Perl Web site and follow the instructions for installation. Perl is an object-oriented procedural language optimized for searching contents, extracting information from arbitrary text files, and printing reports. It is also a good language for many system management tasks. Perl combines (in the author's opinion, anyway) some of the best features of C, sed, awk, and sh, so people familiar with these languages should have little difficulty using it. (We assume you have C experience, so explanations of terminology such as variable, subroutine, and so forth are omitted.) The best way to learn a new programming language is to start with a simple, existing example that closely matches your needs and build upon it—but, of course, you need to learn the basic syntax first. Continuing with our scheme of embedded system design methodology, our strategy is to divide the design requirements into two phases, the user interface and the system interface, and have the database as the center point of interaction between them. Our user interface revolves around Web pages that can be displayed with a browser such as Netscape or Internet Explorer. Perl fits neatly in our scheme, as it provides the CGI module to create dynamic Web pages, and the Pg module to design the database interface. In this chapter, we will develop some Perl scripts as part of the user interface for our embedded system design projects.

A SHORT COURSE ON PERL

Almost every programming book begins by writing the program that prints "Hello World." In keeping with tradition, here is our first Perl program, "Hello World." Create a file "hello.pl" (call it a Perl script) and insert the following text:

```
#!/usr/bin/perl
print "Hello World\n";
```

Change the file permission to make it executable.

```
#chmod 777 hello.pl
```

Execute the script with the following shell command (assuming Perl is already installed on your system):

```
[root@ns /root] #./hello.pl
```

You will see the following response:

```
Hello World
```

A script is a text file that is interpreted by the shell before it is handed over to the actual interpreter. The first line of your Perl file should have the following line:

```
#!/usr/bin/perl
```

It allows the shell to invoke Perl for interpreting the rest of the text.

The comments in Perl start with #. The text following the # is ignored by the interpreter. The output of the following program is same as the one discussed previously:

```
#!/usr/bin/perl
print "Hello World\n"; #My first program
```

Variables

There are three ways to describe a variable in Perl: scalar, array, and hash.

Scalar Variables Start with the "$" Symbol

A single-valued variable is declared by preceding the variable name with a "$". The following line declares a variable "day" without initializing it:

```
$day;      # the simple scalar value "days" without initialization
```

Array Variables Start with the "@" Symbol

A multivalued array variable is declared by preceding the variable name with a "@". The following line declares an array variable "days" without initializing it:

```
@days;      #An empty array
```

Hash (Key and Value Pair) Variables Start with the "%" Symbol

Hash is essentially a dictionary variable that has a key and a value pair associated with each member of the variable. A hash variable is declared by preceding its name with a "%". The following line declares a hash variable "days_are_numbered" without initializing it:

```
%days_are_numbered;       #Hash or keys and value pair
```

Assigning Values to Variables

Variables can be initialized when they are declared; however, they do not have a type associated with them. A variable can store an integer at one point and a string later, depending on the context. Perl converts strings to numbers transparently when required. The following lines declare and initialize variables of scalar, array, and hash with no associated types:

```
$day="Monday";                       # now scalar is a string value,
$day=1;                              # now an integer value
@days=('mon','tue',wed','thu','fri','sat','sun'); # An array of strings
@days=(1,2,3,4,5,6,7);               # An array of integers
```

One part of an array can be used for integers, while the other part can be used for strings.The variable type is relevant only when it is being used in a context. The following array variable is a combination of string and integer values:

```
@dog_days('mon',1,'fri');                 # mix integer and string in assignment
%days_are_numbered=("mon"=>1,"tue"=>2);   # Hash or keys and value pair
```

The names of hash and arrays are interchangeable with scalar. You can extract a single element of a hash or array variable by simply indexing the specific member through bracketing the index as shown here:

```
$day = $days[0];                     # Accessing array element as scalars
$days[0] = $day;                            # Assigning array elements
$day=$days_are_numbered{"mon"};             # Accessing hash element
$days_are_numbered{"wed"} = 3;              # Assigning hash element
$day=$day + 1;                              # Now $day = 2
```

Single- and Double-Quoted Text

There is a difference between text surrounded by single quotes or double quotes. Single-quote text is treated *literally*, while double quotes mean that their contents should be interpreted. For example, the $day will be replaced with the actual value of the variable if surrounded by double quotes, but printed as $day if within single quotes.

```
$day = "tuesday".
print "This is a bad $day.";              #will print Tuesday
print 'This is a bad $day.';              #will print $day
$day = "$day,"               # The $day will be replaced by the value
$day = '$day,'               # The $day will not be replaced by its value
\n                           # is a new line
$day="\\"                    # is a \
```

(Two other useful backslash sequences are \t to insert a tab character, and \\ to insert a backslash into a double-quoted string.)

Special Properties of Variables

Perl interprets the #, %, and $ symbols in a special way when associated with a variable name; for example, a $ sign followed by # in the name of an array returns the last element of the array. The $% or $$ returns the current process ID as shown here:

```
$#days                       # the last index of array @days
$% or $$.                    # is the current process id.
```

Range of Numbers

A shortcut to specify a range of numbers in a loopstructure is by using "..". For example, you can write (1, 2, 3, 4, 5) as (1 .. 5). You can also use arrays and scalars in your loop list. The following code segment prints numbers 1 through 10, 15, and then 20 through 25:

```
@one_to_ten = (1 .. 10);
$top_limit = 25;
for $i (@one_to_ten, 15, 20 .. $top_limit)
{
    print "$i\n";
}
```

Special Variables

Following are some of the more commonly used special variables (for a complete listing, refer to the Perl man pages). The explanation is given following the comments on each line

$! Contains the most recent warning generated by Perl.

%ENV Contains the script environment variable. The following example will print all env variables:

```
foreach (keys %ENV)
{
print "$_=$ENV{$_}\n";
}
```

$_ The default scalar variable for passing parameters, and input and pattern-searching space.

The following pairs are equivalent:

```
/^Subject:/
$_ =~ /^Subject:/
tr/a-z/A-Z/
$_ =~ tr/a-z/A-Z/
chomp
chomp($_)
print()
print($_)
```

$<digits> Contains the subpattern from the corresponding set of capturing parentheses from the last pattern match.

$~ The name of the current report format for the currently selected output channel. The default is the name of the filehandle.

$! If used numerically, yields the current value of the C "errno" variable. If used as a string, yields the corresponding system error string.

$$ The process number of the Perl running this script.

$> The effective uid of this process.

$(The real gid of this process.

$) The effective gid of this process.

$0 Contains the name of the program being executed.

@ARGV The array @ARGV contains the command-line arguments intended for the script. "$#ARGV" is generally the number of arguments minus 1, because "$ARGV[0]" is the first argument, not the program's command name itself.

@_ Within a subroutine, the array @_ contains the parameters passed to that subroutine.

%SIG The hash %SIG contains the signal handlers for signals.

$SIG{expr} As part of the %SIG hash, the $SIG points to function that is supposed to handle the signal.

For example:

```
sub handler
{ # 1st argument is signal name
    my($sig) = @_;
     print "Caught a SIG$sig—shutting down\n";
     exit(0);
}
$SIG{'INT'}= \&handler;
$SIG{'QUIT'} = \&handler;
```

Everything is OK when assigning a value to variables. As explained earlier, there is no specific type associated with the value of a variable. For example, assigning 1 to a variable $day can be treated as a string, or can be used in an expression for evaluation.

```
$Day=1              # Variables do not have types associated with them
$Day=1.23           # It is a floating point number
$Day=2E10           # A very large number
$Day=0xfff          # A hex number
$Day=0377           # An octal number
$Day=4_294_1        # underline for legibility
@days[3,4,5]        # same as @days[3..5]
@days{'a','c'}      # same as ($days{'a'},$days{'c'})
%days{"Ugh", "monday","Ahh","saturday"}
```

It is often more readable to use the => operator.

```
%days{"Bad",=>"monday","good"=>"saturday"}
```

This is OK, too:

```
%map = (    # This is OK too.
    red => 0x00f,
    blue => 0x0f0,
    green => 0xf00,
);
$rec = {# initializing hash references to be used as records
    who => 'let the dog out',
    cat => 'on the hot tin roof'
};
```

FUNCTIONS AND SUBROUTINES

Perl is a procedural language similar to C and C++. It uses the concept of subroutines as in C to block a section of code. The subroutines can be used repeatedly by simply calling their names in a different code block with an initial "&" preceded with the name, but it is optional. The following two lines have identical meanings for declaring subroutines:

```
&HelloWorld;
HelloWorld();               #This is OK too
```

Defining subroutines require the word "sub" preceded with the name of the subroutine as shown here:

```
sub HelloWorld
{
print "Hello","World";      # perl will combine Hello and World
}
```

The arguments passed into the functions are in the form of a single list of scalars. If multiple lists are passed, they are concatenated into one.

```
@try1=("this","that");
@try2=("he","she");
trying(@try1,@try2);
trying("this","that","he","she");         #same as above
```

You can declare a subroutine without defining it by saying "sub name"; for example:

```
sub myname;
```

If you do not expressly use the return statement, the sub returns the result of the last statement.

LOCAL VARIABLES

$my creates a local variable that is known only inside the sub.

```
$day="mon"
sub HelloWorld {
$my day="tue";
print $day;                # will print tue.
}
```

PERL STATEMENTS

A Perl script consists of a sequence of declarations and statements. The only things that need to be declared in Perl are report formats and subroutines. The sequence of statements is executed just once.

Simple Statements

Expressions are simple statements that are evaluated with the rules defined similar to the C language expression evaluation. Every simple statement must be terminated with a semicolon, unless it is the final statement in a block, in which case, the semicolon is optional.

```
$day = 5;
$day = $day + 15;
```

File Handling

filehandles are special variables that are created when a file is open for reading and writing. Conventionally, the name of the filehandle is written in uppercase. The less than and greater than signs are used before the filename to associated a file for reading, writing, or append mode of operation, as shown here:

```
open (MYFILE, "File name");       #opens a file for reading
open (MYFILE, "<File name");      #opens a file for reading
open (MYFILE, ">File name");      #opens a file for writing
open (MYFILE, ">File name");      #opens a file for appending
```

Opening a file for write only will overwrite the existing data. If you want to preserve the existing data, open a file in append mode as shown in line 4 in the preceding code. To read from a file one line at a time, use the following syntax:

```
$line=<MYFILE>;
chop $line; # newline character is being removed
```

To read the entire file at once, use the array variable.

```
@allline=<MYFILE>;
```

To read the entire contents of a directory filename, use file globing with the wildcard character *. The following statement reads all the filenames with a .pl extension:

```
@allfiles=<*.pl>;
```

Write to a file is done with the print statement.

```
print MYFILE $line;
```

If no filehandle is given, then standard output is being used to print out.

```
# Read a text file and print it out
while (<$filename>)
{
    print;
}
```

The default filehandle is STDIN and does not need to be open exclusively; to read from STDIN, simply use an empty angle symbol <>.

```
    $line = <>;    #read one line from standard input
    print $line
# Copy a binary file to another file
open (OUTFILE,“>>/tmp/myfile”);
while ($bytesread=read($filename,$buffer,1024))
{
            print OUTFILE $buffer;
}
```

It is always possible for an open() call to fail. In Perl, you can guard against these problems by using “or” and “and.” A series of statements separated by “or” will continue until you hit one that works, or returns a true value. This line of code will either succeed at opening OUTPUT in overwrite mode, or cause Perl to quit:

```
open (OVERWRITE, ">overwrite.txt") or die "$! error trying to overwrite";
```

Reserved Functions

split()

This function breaks apart a string and returns a list of the pieces. split() generally takes two parameters: a regular expression to split the string with, and the string. You can also specify a third parameter: the maximum number of items to put in your list. The splitting will stop as soon as your list contains that many items. The following program demonstrates the use of the split function. The result is an array @a with two items, "Hello." and "Welcome Perl!\n".

```
1) $line=<>;                    #opens line from std input
2) chop $line;                  #removes newline
3) @alllines=<FILEHANDLE>       #reads all the lines in the file
4) @allfiles=<*.*>              #reads all files in the directory
5) $a+=1                        # same as $a = $a + 1;
6) $a = "Welcome to Perl!\n";
7) print substr($a, 0, 7);      # "Welcome"
8) rint substr($a, 7);     # " to Perl!\n"
9) print substr($a, -6, 4);     # "Perl"
10) $a = "Hello. Welcome Perl!\n";
11) @a = split(/ /, $a);        # split along spaces
12) # Three items: "Hello.", "Welcome", "Perl!\n"
```

join()

The opposite of split is join(). The join() function takes a list of strings and attaches them together with a specified string between each element, which can be an empty string.

```
@a = ("Hello.", "Welcome", "Perl!\n");
$a = join(' ', @a);        # "Hello. Welcome Perl!\n";
$b = join(' and ', @a);    # "Hello. and Welcome and Perl!\n";
$c = join('', @a);         # "Hello.WelcomePerl!\n";
```

Conditional Statements

You can use any of the following symbols for evaluation of an expression for a true or false result:

```
>        gt      # greater than
<        lt      # less than
>=       ge      # greater than or equal to
<        lt      # less than or equal to
==       eq      # equal to
!=       ne      # not equal to
<=>      cmp     # comparison
```

if, and if than else

```
if ($day == 5)
{
    print "5 days a week";
}
else
{
    print "10 days a week";
};
```

The preceding if than else is a tertiary conditional statement.

```
($day == 5) ? print "5 days a week":  print "10 days a week";
```

Loop Control Operators

These operators allow repeated execution of a block of code. The repeat count can be a predetermined number, or a conditional statement can be included to determine the condition for terminating the loop.

One-Line Loop Construct

You do not need to enclose the loop constructs in braces if you have a short statement block that can be placed on one line. The following "while" loop demonstrates a single control statement without braces:

```
$a+=1    while    $a<=10;
$a+=1    until    $a!=10;
```

Multiple-Line Loop Construct

Use *while* to loop over a block of code surrounded by braces.

```
while ($a <= 10)
{
    print ":$a:";
```

```
}
while (<MYFILE)                    #until the end of file
{
   push @filecontents, $_;         #the array is filled with file lines
}
```

Use *for* to loop over a block of code, surrounded by braces.

```
for ($a = 0; $a <= 10; $a++)       #start a=0, terminator a <=10, inc a
{
   print ":$a:";
}
```

Use *foreach* to loop over a block of code; each time, take a list in parentheses as an argument.

```
@days=("mon","tue","wed");
foreach $a (@days)
{
   print ":$a:";                   # print :mon::tue::wed:
}
foreach (@days)
{
   print ":$_:";                   # print :mon::tue::wed:
}
```

You can put *labels* inside a block of loop construct with ":"

```
foreach (@days)                    # if days=("mon","tue","wed")
{
TEST:
   print ":$_:";                   # print :mon::tue::wed:
}
```

Use *redo* to repeat without evaluating the condition.

```
foreach (@days)                    # if days=("mon","tue","wed")
{
redo if ($day[0] == "mon");               # loop infinite
   print ":$_:";                   # print :mon::tue::wed:
}
```

Use *next* with a label to skip the remaining code and begin with evaluating the variable.

```
THIS: foreach (@days)                       # if days=("mon","tue","wed")
{
next THIS  if ($day[0] == "mon");               # skip mon
    print ":$_:";                           # print :tue::wed:
}
```

Use *last* with a label to skip the remaining code and begin with evaluating the variable.

```
foreach (@days)                             # if days=("mon","tue","wed")
{
last THIS  if ($day[0] == "mon");               # skip mon
    print ":$_:";                           # print :tue::wed:
THIS:
}
```

BASIC BLOCKS AND SWITCH STATEMENTS

There is no official switch statement in Perl. The BLOCK construct is particularly suitable for doing case structures.

```
SWITCH: {
if (/^abc/) { $abc = 1; last SWITCH; }
if (/^def/) { $def = 1; last SWITCH; }
if (/^xyz/) { $xyz = 1; last SWITCH; }
$nothing = 1;
}
```

Pattern Matching and Substitution

Perl offers comprehensive pattern-matching and substitution capabilities. In its simplest form, you can specify a given pattern, and Perl returns true or false. For substitution, specify a pattern, and Perl will substitute the string. The $_ variable is used as the default for both matching and substitution.

```
$_ = $Testing;
/PATTERN/;
#same as
$Testing =~/PATTERN/;
print "$Testing contains PATTERN" if ($Testing=~/PATTERN/);
```

You can perform matching only for testing purposes, such as:

```
m/PATTERN/;                          #matching operator
```

or substitute an occurrence of a string with a replacement string, such as:

```
s/PATTERNString/REPLACEString/;   #substitution operator
```

When you want to use a regular expression to match against a string, you use the special =~ operator.

```
$a = "Hello. Welcome Perl!\n";
if ($string =~ /Welcome/)
{
    print "Have Fun!\n";
}
```

The syntax of a regular expression for matching is a string within a pair of slashes. The code $string =~ /welcome/ asks whether the literal string welcome occurs anywhere inside $string. If it does, then the test evaluates true; otherwise, it evaluates false.

```
m!PATTERN!;                #matching operator can have !
```

The pattern inside the matching operator behaves similarly to a string enclosed in double quotes, so values replace variables.

```
$_ = "How are you";
$Testing = "How";
/$Testing/;
print $_;  # will print the number of times "How" is being found in $_
```

Metacharacters

Introducing some special characters can modify the behavior of pattern search within the enclosed slashes of the pattern search string. These special characters are called *metacharacters*.

The two simplest metacharacters are ^ and $. These indicate "beginning of string" and "end of string," respectively.

Finding an Occurrence of a Pattern at the Beginning of a String Using ^

Inserting the ^ character following the beginning slash starts a pattern search that shows up only at the beginning of the string. For example, the pattern search with

/^Tom/ will match "Tom was here", "Tom" and "Tommy". But it won't match "Here is Tom!" or "This is Tom", because Tom doesn't appear at the beginning of the string.

Finding an Occurrence of a Pattern at the End of a String Using $

The $ character provides a matching of the pattern that shows up at the end of the string. The expression /dogs$/ will match "cats and dogs," but not "dogs and cats."

The following code segment demonstrates a pattern search that identifies whether a valid URL is present in a string. If the line starts with http: and ends with html, it is a potential URL.

Here is a simple routine that will take lines from a file and only print URLs that seem to indicate HTML files:

```
for $line (<MYFILE>)
{
    if (($line =~ /^http:/) and ($line =~ /html$/))
    {
    print $line;
    }
}
```

Wildcard Substitution with + .? *

The + metacharacter allows a substitution one or more times. The regular expression /ab+c/ will match "abc," "abbc,'" "abbbc," and so on.

A period inside a regular expression will match any character, except a newline. For example, regexp /a.b/ will match anything that contains another character that is not a newline, followed by a, such as "aab," "a3b," "a b," and so forth.

The * quantifier matches the immediately preceding character or metacharacter zero or more times. This is different from the + quantifier. /ab*c/ will match "abc," "abbc," and so on, just as /ab+c/ did, but it will also match "ac," because there are zero occurrences of b in that string. The quantifier will always match as many characters as it can.

```
$book_pref = "The cat in the hat is where it's at.\n";
$book_pref =~ /(cat.*at)/;
print $1, "\n";
```

The matching expression (cat.*at) is greedy. It contains "The cat in the hat is where it's at" because that is the largest string that matches.

Finally, the ? quantifier will match the preceding character zero or one times. The regex /ab?c/ will match "ac" (zero occurrences of b), and "abc" (one occurrence of b). It will not match "abbc," "abbbc," and so on.

Searching for Metacharacters Literally

In order to search for the occurrence of any of the metacharacters "+ .? *", you must precede it with a backslash, The regex /Mr./ matches anything that contains "Mr" followed by another character. If you only want to match a string that actually contains "Mr.", you must use /Mr\./.

Matching Digits and Words

Perl provides special matching metacharacters that match only specific types of characters, such as:

\d will match a single digit.

\w will match any single "word" that begins with a letter, digit, or underscore.

\s matches a whitespace character, including space, as well as the carriage return linefeed.

These metacharacters work like any other character; you can match against them, or you can use quantifiers such as + and *. The regex /^\s+/ will match any string that begins with whitespace, and /\w+/ will match a string that contains at least one word.

One good use for \d is testing strings to see whether they contain numbers. For example, you might need to verify that a string contains an American-style telephone number with the form 555-1212. You could use code like this:

```
unless ($phone =~ /\d\d\d-\d\d\d\d/)
{
   print "That's not a phone number!\n";
}
```

Specifying Repeat Counts

You do not have to repeat the \d multiple times; you can specify the repeat count within curly braces as shown here:

```
unless ($phone =~ /\d{3}-\d{4}/)
{
   print "That's not a phone number!\n";
}
```

The string \d{3} means to match exactly three numbers, and \d{4} matches exactly four digits. If you want to use a range of numbers, you can separate them with a comma; leaving out the second number makes the range open-ended. \d{2,5} will match two to five digits, and \w{3,} will match a word that is at least three characters long.

Excluding words and digits from the search criteria: You can invert the \d, \s, and \w metacharacters to refer to anything but that type of character. \D matches nondigits; \W matches any character that is not a word, meaning not a letter, digit, or underscore; and \S matches anything that is not a carriage return, linefeed, space, or tab.

Defining your own character class: If you want to restrict your search to a specific set of characters, you can define them by enclosing a list of the allowable characters in square brackets. For example, a class containing only the lowercase vowels is [aeiou]. /b[aeiou]g/ will match any string that contains "Ka", "Ki", "Ku", "Ke", or "Ko". For specifying a range of characters, use dashes; for example, [a-f]. You can combine character classes with quantifiers:

```
if ($string =~ /[aeiou]{2}/)
{
    print "This string contains at least two vowels in a row.\n";
}
```

Excluding character class: You can also invert character classes by beginning them with the ^ character. An inverted character class will match anything you do not list. [^aeiou] matches every character except lowercase vowels. Notice that the ^ is also being used for matching the beginning of a string; the difference is the enclose slash for matching the beginning of a string.

i Flag

Case-insensitive: If you want to ignore case in your search pattern, use the i flag, which makes a match:

```
$sentence = "Thank you!";
if ($sentence =~ /thank/i)
{
    print "you are welcome!\n";
}
```

g Flag

Repeated search: Searching for multiple occurrences is accomplished by using *g* flag with the regular expression. It will tell Perl to remember where it was in the string when it returns to it.

```
$number = "He is 6ft tall and weighs 200lb.";
while ($number =~ /(\d+)/g)
{
    print "found the number $1.\n";
    $count++;
}
print "There are $count numbers here.\n";
```

Subexpressions

Perl offers a convenient way to search more than one item in a text string using the special character |. The following example illustrates a situation in which the contents of a file are searched for the words *Alabama*, *Alaska*, *Arkansas*, and *Arizona*. The match will return a true if the substring is present in the text.

```
@address_lines = ("1111 Main St. Phoenix, AZ. 99999");
for $check (@address_line)
{
        if ($check =~  /Alabama|Alaska|Arkansas|Arizona/)
        {
        print "Your address is $check"\n";
        }
}
```

The string that caused the subexpression to match will be stored in the special variable $1. We can use this to extract the matching part only:

```
for $check (@address_line)
{
    if ($check =~  /Alabama|Alaska|Arkansas|Arizona/)
    {
    print "You live in state of $1\n";
    }
}
```

Nesting of Subexpressions

If your expression contains more than one subexpression, the results will be stored in variables named $1, $2, $3, and so on. This can be very handy in situations where you need to extract the occurrence of each subexpression, such as retrieving the full time, hours, minutes, and seconds separately from a string that contains a timestamp in hh:mm:ss format.

```
$string = "The time is 9:30:30.";
$string =~ /((\d{1,2}):(\d{2}):(\d{2}))/;
@time = ($1, $2, $3, $4);
($time, $hours, $minutes, $seconds) = ($1, $2, $3, $4);
```

The expression \d{1,2} will match single or double digits, such as 9 or 12. You can directly assign the values to a list of variables as shown here:

```
($time, $hours, $minutes, $seconds) =
($string =~ /((\d{1,2}):(\d{2}):(\d{2}))/);
```

Search and Replace

Perl can substitute the occurrence of a string with a given string just as we do in our word processor search and replacement function. The syntax for search and replacement is:

```
s/search/replace/
```

The s/// operator searches for the given string and replaces it with the replacement string. For example, the following code will change Bob to Harry:

```
$greet = "Hi! Bob\n";
$greet =~ s/Bob/Harry/;
print $greet;
```

You can also use subexpressions in your matching expression, and use the variables $1, $2, and so on that they create. The replacement string will substitute these, or any other variables, as if it were a double-quoted string. The following code segment demonstrates the use of the $1 and $2 variables that were created while searching for a given string. The sentence "me and you" is being replaced with "you and me."

```
$greet = "me and you\n";
$name =~ s/(\w+),(\w+),(\w+)/$3 $2 $1/;  # "You and me"
```

The built-in functionality of search and replacement is a very powerful tool that can ease the burden of writing complex code for any Web application. Next, we will discuss the object-oriented capability of Perl through integrating Perl modules in the design.

PERL MODULES

Perl modules make Perl programming a pleasant experience, especially the CGI and Pg modules. The object-oriented interface allows you to build complex applications just by incorporating few basic functions of these modules into your CGI scripts. The CGI is the most commonly used method of providing dynamic interface through a Web client and a Web server. The Internet has truly standardized the look and feel of the user interface. Every form on the Internet has a text box, a check box, a radio button, pop-up menus, scrolling lists, and a few buttons, or a combination thereof. The CGI.pm module has built-in functions that essentially transform the script programming into a simple job of filling out templates or simple one-line function calls.

The Pg module, on the other hand, provides the object-oriented interface to the database server PostgreSQL. Looking back to our methodology for the user interface, we need the CGI module to create the desired HTML output, and the Pg module to interface with the database. Several examples in Chapter 9, "Embedded System Design Projects," incorporate the two modules in the Web scripts and generate the user interface as part of the system design. Besides CGI and Pg, the CPAN is full of Perl modules for many other programming needs. Therefore, if you think you have a specific programming requirement, refer to CPAN first before venturing into writing a module yourself. If your Linux distribution does not provide you with a Perl module, you can simply download it from the following Web site:

```
http://www.genome.wi.mit.edu/ftp/pub/software/WWW/cgi_docs.html
ftp://ftp-genome.wi.mit.edu/pub/software/WWW/
```

Next, we discuss the two modules in greater detail, because we will be using them in our scripts for generating the user interface.

CGI.pm

Creating Web pages is about creating HTML tags. CGI is essentially a server mechanism of spawning a shell and redirecting the output stream to the client rather than STDOUT. The variables received from the client through the GET and POST methods are passed along to the shell as environment variables. The shell executes the required script, and the output generated by the script is sent back to the client via a TCP/IP link. The CGI module provides a function-oriented interface, as well as an object-oriented interface that receives the environment variables and makes them available as any other variables for the programming interface. To include CGI as the *function-oriented interface*, simply add the module to the namespace, and use the functions listed in the module as shown in the following code segment:

```
use CGI qw/:standard/;
print header,
start_html('Hello World'),
h1(' Hello World '),
end_html
```

The qw/:standard/ specify import tag and the function start_html and h1 are part of the CGI module and are included in the namespace.

The *object-oriented interface* requires you to create an instance of the object using the new() method. With this scheme, you can examine keywords and parameters passed to your script, and create forms whose initial values are taken from the current query, thus preserving the state information.

```
#!/usr/bin/perl -w
use CGI;                          # load CGI routines
$query = new CGI;                 # create new CGI object
print $query->header,                    # create the HTTP header
$query->start_html('hello world'), # start the HTML
$query->h1('hello world'),        # level 1 header
$query->end_html;                 # end the HTML
```

This will parse the input (from both POST and GET methods) and store it into the object called $query.

Next, we will discuss ome important functions within the CGI module that help create dynamic HTML pages to be used in the user interface part of our embedded system designs.

CGI Functions

The CGI provides comprehensive functionality to control the behavior of the document and produce HTML output that meets a specific requirement. The following are common groups of functions available for creating HTML documents.

HTTP header()

The first step in HTML pages is creating the header. This tells the browser what type of document to expect, and gives other optional information, such as the language, expiration date, and whether to cache the document. The header can also inform the browser whether to leave the connection open with server push. The following is a simple header:

```
print header;
```

or

```
print $query->header;
```

This will produce the output:

```
"Content-type: text/html\n\n"
```

You can provide your own MIME type if you wish; otherwise, it defaults to text/html. You can pass arguments such as:

```
print $q->header(-type=>'image/gif',-expires=>'+3d');
```

or

```
print $query->header('image/gif');
```

or

```
print $query->header('text/html','204 No response');
```

If you specify an absolute or relative expiration interval, some browsers and proxy servers will cache the script's output until the indicated expiration date. You can specify expiration in the following format with -expires:

+30s	30 seconds from now
+10m	ten minutes from now
+1h	one hour from now
-1d	yesterday (i.e., "ASAP!")
now	immediately
+2M	in two months

+1y	in one year time
25-June-2002 00:50:33	at the indicated time & date

```
print $query->header(-type=>'image/gif', -nph=>1, -status=>'402 Payment required',
            -expires=>'+3d',
            -cookie=>$cookie,
            -Cost=>'$2.00');
```

Each argument name is preceded by a dash. The argument names are case insensitive; -type, -Type, and -TYPE are all acceptable. The order of the arguments is also not important. If a routine is being called with just one argument, you can omit the name; for example:

```
print $q->header('text/html');
```

In this case, the single argument is the document type.

Perl does not waste any named arguments that are not being recognized. For example, the following header will still create some output:

```
print $q->header(-type         =>  'text/html',
                 -cost         =>  'Pending',
                 -new_buyers   =>  'many');
```

This will produce the following nonstandard HTTP header:

```
HTTP/1.0 200 OK
Cost: Pending
New-Buyers: many
Content-type: text/html
```

Note that the underscores are translated automatically into hyphens.

If you provide a filehandle to the new method, it will read parameters from the file (or STDIN); for example:

```
$query = new CGI(INPUTFILE);
```

You can use curly braces when calling any routine that takes named arguments, as shown here:

```
print $q->header( {-type=>'image/gif',-expires=>'+3d'} );
```

The cookie parameter generates a header that tells the browser to provide a "magic cookie" during all subsequent transactions with your script. Netscape cookies have a special format that includes interesting attributes such as expiration time. Use the cookie() method to create and retrieve session cookies.

Redirection header()

If you do not want to return a document to the browser, you have the option of pointing it to another URL:

```
print $query->redirect('http://www.aklinux.com');
```

or

```
print redirect('http://www.perl.org/');
```

Script Initialization

When a user receives a new fill-out form the first time, the parameters are empty, but when a form is being filled out, the same query has values. The default values that you specify for the forms are only used the first time the script is invoked (when there is no query string). On subsequent invocations of the script (when there is a query string), the former values are used even if they are blank. If you want to change the value of a field to its previous value, you can either call the param() method to set it, or use the -override (alias -force) parameter.

You can check to see if the form is being filled by calling the param() routine, as shown in the following listing:

```
use CGI qw/:standard/;
print header,
    start_html('A Simple Example'),
    h1('A Simple Example'),
    start_form,
    "Name ",textfield('name'),p,
    "Status ", p,
    checkbox_group(-name=>'status',
                   -values=>['Active','Passive'],
                   -defaults=>['Active']), p,
    popup_menu(-name=>'state',
               -values=>['AL','AK','AZ']),p,
    submit,
    end_form,
    hr;
```

```
if (param()) {
    print "The name: ",em(param('name')),p,
          "The State: ",em(param('state'))),p,
          "Status: ",em(param('status')), hr;
}
```

You can initialize the routine with any type of argument, be it array or scalar, and the routine will do whatever is most appropriate. For example, the param() routine is used to set a CGI parameter to a single or a multivalued value, as shown here:

```
$q->param(-status=>'state',-value=>'CA');
$q->param(-name=>'state',-value=>['AL','AK','AZ']);
```

To create an empty query, initialize it from an empty string or hash:

```
$empty_query = new CGI("");
```

or

```
$empty_query = new CGI({});
```

Fetching the Names of All the Parameters Passed to Your Script

If the script was invoked with a parameter list (e.g., "&name1=value1 &name2=value2 &name3=value3"), the param() method will return the parameter names as a list.

```
@names = $query->param
```

Fetching the Value of a Single Named Parameter

Pass the param() method a single argument to fetch the value of the named parameter. If the parameter is multivalued (e.g., from multiple selections in a scrolling list), you can ask to receive an array. Otherwise, the method will return a single value.

```
@values = $query->param('test');
```

or

```
$value = $query->param('test');
```

Fetching the Parameter List as a Hash

The Vars() method can be used to fetch the entire parameter list as a hash in which the keys are the names of the CGI parameters, and the values are the parameters' values. There is a difference if the function is being asked to return a reference; for example:

```
$params = $q->Vars;
print $params->{'address'};
@foo = split("\0",$params->{'foo'});
```

In which case, changing a key changes the value of the parameter in the underlying CGI parameter list. If asked to return in an array context, it returns the parameter list as an ordinary hash. This allows you to read the contents of the parameter list, but not change it.

```
%params = $q->Vars;
```

Deleting a Parameter

This completely clears a parameter. It sometimes useful for resetting parameters that you do not want passed down between script invocations. If you are using the function call interface, use "Delete()" instead to avoid conflicts with Perl's built-in delete operator.

```
$query->delete('foo');
```

Use delete_all() to delete all parameters.

HTML Start Document

After the header comes the document. This is where the opening tags of <HEAD> and <TITLE> and the <BODY> are produced. The following is a simple document head, title, and body:

```
print start_html;
```

You can specify optional information that controls the page's appearance and behavior in start_html. All parameters are optional. In the named parameter form, recognized parameters are -title, -author, -base, -xbase, and –target.

The argument -xbase allows you to provide an HREF for the <BASE> tag different from the current location. All relative links will be interpreted relative to this tag, as in, <BASE HREF="http://www.aklinux.com/UCIrvine/">.

The argument -target allows you to provide a default target frame for all the links and fill-out forms on the page.

```
print $query->start_html(-title=>'Floating point arithmetic',
                         -author=>'akhan@aklinux.com',
                         -base=>'true',
                         -target=>'fixed_point');
```

HTML End Document

The last thing in the document is the end_html. This ends an HTML document by printing the </BODY></HTML> tags.

```
print $query->end_html
```

HTML Elements

CGI offers a simplified scheme of generating HTML tags that eliminates typing errors. There are no specific functions, but the tags are created dynamically as needed. These are the "HTML shortcuts":

```
print p ('testing');      #this will produce the paragraph
print h1 ('testing');     #this will headline
```

HTML tags have both attributes (the attribute="value" pairs within the tag itself) and contents (the part between the opening and closing pairs). To distinguish between attributes and contents, CGI.pm uses the convention of passing HTML attributes as a hash reference as the first argument, and the contents, if any, as any subsequent arguments. It works out like this:

Code-Generated HTML	
h1()	<H1>
h1('Bold','characters');	<H1>Bold characters</H1>
h1({-align=>left});	<H1 ALIGN="LEFT">
h1({-align=>left},'contents');	<H1 ALIGN="LEFT">contents</H1>

Form Elements

One of the most popular uses of Perl CGI is generating forms over the Internet. There is a simple correlation between HTML form tags and CGI's way of generating the HTML output as shown previously. The HTML forms essentially consist of three elements: form headers, one or more input fields, and buttons. CGI supports different field types, such as single or multiple line input, password fields, hidden fields, list boxes, and so forth, as well several different button types suitable for a variety of form applications. The following are the general form elements supported by the CGI form module.

Starting and Ending a Form

The HTML tag for form is <FORM>. The Perl CGI invocation of form is startform(). It will return a <FORM> tag. The following function returns a <FORM> tag:

```
print $query->endform;
```

with the default action, method, and enctype as shown here:

```
method: POST
action: this script
enctype: application/x-www-form-urlencoded
```

You can provide optional parameters of action, method, and enctype in the function call:

```
print $query->startform(-method=>$method,
-action=>$action,
-enctype=>$encoding);
```

or

```
print $query->startform($method,$action,$encoding);
```

The endform() returns the closing </FORM> tag.

```
Print end_form();
```

Arguments

The enctype argument tells the browser how to package the various fields of the form before sending the form to the server. The two possible values are application/x-www-form-urlencoded. This is the older type of encoding used by all browsers prior to Netscape 2.0, and the newer multipart/form-data type that allows file uploading and supports very large fields.

The CGI.pm stores the name of this encoding type in &CGI::MULTIPART. The startform() uses the default older method; however, you can use start_multipart_form() instead, to include the newer multipart/form-data type.

ISINDEX Field

Ideal for sites that provide search engines. The tag inserts an input field into the document so the visitors can enter search queries. The queries are transferred to the CGI application indicated by the URL specified by Action=attribute. The default is to process the query with the current script.

```
print $query->isindex(-action=>$action);
```

or

```
print $query->isindex($action);
```

The equivalent HTML tag and attribute syntax is:

```
<ISINDEX ACTION="/cgi-bin/search.cgi">
```

Fetching a List of Keywords from the Query

If the script was invoked as the result of an <ISINDEX> search, the parsed keywords can be obtained as an array using the keywords() method.

```
@keywords = $query->keywords
```

Text Fields

A single-line text input window is being shown on the browser for the user to enter the appropriate information:

```
print $q->textfields(-name=>'LastName', -default=>'Enter Name', -size=>30, -
maxlength=>30);
```

The equivalent HTML tag and attribute syntax is:

```
<INPUT TYPE=text NAME="LastName" SIZE=30 MAXLENGTH=32 VALUE="Enter Name">
```

Parameters

name (required) Name for the field first parameter.

default (optional) The second parameter is the default starting value for the field contents.

size (optional) The third parameter is the size of the field in characters.

maxlength (optional) The fourth parameter is the maximum number of characters the field will accept.

The fields will be initialized with the previous contents from earlier invocations of the script. When the form is processed, the value of the text field can be retrieved with:

```
$value = $query->param('LastName');
```

Multiple-Line Text Areas

The function is similar to the single-line text input window, except that the user can input multiple line of text data.

```
print $q->textarea(-name=>'Message', -default=>'Enter a message', -rows=>10, -
cols=>40);
```

or

```
print $query->textarea('Message','starting value',10,50);
```

The equivalent HTML tag and attribute syntax is:

```
<TEXTAREA NAME="Message" ROWS=10 COLS=40 > Enter a message </TEXTAREA>
```

Password Field

The password field is similar to the single-line text input window, except that the input characters are hidden for security purposes and only asterisks are printed.

```
print $q->password_fields(-name=>'Password', -default=>'', -size=>6, -
maxlength=>6);
```

The equivalent HTML tag and attribute syntax is:

```
<INPUT TYPE=password NAME=password SIZE=6 MAXLENGTH=6>
```

Standalone Check Box

The standalone check box provides single check selection for user entry, such as "do you wish to be included in mailing list?"

```
print $q->checkbox(-name=>'mailinglist', -value=>'yes', checked=>'true');
```

The equivalent HTML tag and attribute syntax is:

```
<INPUT TYPE=checkbox NAME='mailinglist' VALUE='yes' CHECKED='true'>
```

Check Box Group

The check box group is similar to radio buttons, but allows more than one choice for input; for example, days of the week a newspaper is to be delivered.

```
%days_of_week= (mon=>1,tue=>2,wed=>3,thu=>4,fri=>5,sat=>6,sun=>7);
print $q->days_of_week(-name=>'days', -values=>['sat', 'sun'], -
default=>'male', -linebreak=>'true', -labels=>\%days_of_week);
```

The equivalent HTML tag and attribute syntax is:

```
<INPUT TYPE=radio NAME='sex' VALUE='male' CHECKED>
<INPUT TYPE=radio NAME='sex' VALUE='female'>
```

Arguments

name and values (required) The first and second arguments are the check box name and values, respectively. As in the pop-up menu, the second argument should be an array reference. These values are used for the user-readable labels printed next to the check boxes, as well as for the values passed to your script in the query string.

default (optional) The third argument can be either a reference to a list containing the values to be checked by default, or a single value to be checked. If this argument is missing or undefined, nothing is selected when the list first appears.

linebreak (optional) The fourth argument can be set to true to place line breaks between the check boxes so that they appear as a vertical list. Otherwise, they will be strung together on a horizontal line.

labels (optional) The fifth argument is a pointer to an associative array relating the check box values to the user-visible labels that will be printed next to them. If not provided, the values will be used as the default.

File Upload

The filefield() will return a file upload field. You must use the new multipart encoding scheme for the form for this method. You can do this either by calling startform() with an encoding type of $CGI::MULTIPART, or by calling the new method start_multipart_form() instead of the old start_form().

```
print $query->filefield(-name=>'uploaded_file',
                        -default=>'starting value',
                        -size=>50,
                        -maxlength=>80);
```

or

```
print $query->filefield('uploaded_file','starting value',50,80);
```

When the form is processed, you can retrieve the entered filename by calling param():

```
$filename = $query->param('uploaded_file');
```

Arguments

name (required) The name for the field's first parameter.

default (optional) The second parameter default has no effect.

size (optional) The third parameter is the size of the field in characters.

maxlength (optional) The fourth parameter is the maximum number of characters the field will accept.

Pop-up Menu

The popup_menu() will return a selectable menu. You must use the new multipart encoding scheme for the form for this method.

```
print $query->popup_menu('states_name',
    ['FL','TX','AZ'],'FL');
```

or

```
%labels = ('FL'=>'Sunshine State','TX'=>'loanstar state',
    'AZ'=>'desert state');
print $query->popup_menu('state_name',
['CA','TX','AZ'],'CA',\%labels);
```

or (named parameter style)

```
print $query->popup_menu(-name=>'state_name',
                         -values=>['FL','TX','AZ']
                         -default=>'FL',
                         -labels=>\%labels);
```

When the form is processed, the selected value of the pop-up menu can be retrieved using:

```
$popup_menu_value = $query->param('state_name');
```

Arguments

name (required) The name for the field, first parameter.

values An array referencing the list of menu items in the menu. You can pass a named array as a reference.

default (optional) The third parameter is the name of the default menu choice. If not specified, the first item will be the default. The values of the previous choice will be maintained across queries.

labels (optional) The fourth argument is a pointer to an associative array relating the check box values to the user-visible labels that will be printed next to them. If not provided, the values will be used as the default.

Scrolling List

The scrolling_list() method will generate a scrolling list form element.

```
print $query->scrolling_list(-name=>'list_name',
                             -values=>['CA','TX','AZ'] ,
                             -size=>3,
                             -multiple=>'true',
                             -labels=>\%labels);
```

or

```
print $query->scrolling_list('list_name',
    ['CA','TX','AZ'] ,3,'true',\%labels);
```

When this form is processed, all selected list items will be returned as a list under the parameter name "list_name." The values of the selected items can be retrieved with:

```
@selected = $query->param('list_name');
```

Arguments

name and values (required) The first and second arguments are the check box name and values, respectively. As in the pop-up menu, the second argument should be an array reference. These values are used for the user-readable labels printed next to the check boxes, as well as for the values passed to your script in the query string.

default (optional) The third argument can be either a reference to a list containing the values to be checked by default, or a single value to be checked. If this argument is missing or undefined, then nothing is selected when the list first appears.

size (optional) The fourth argument is the size of the list.

multiple (optional) The fifth argument can be set to true to allow multiple simultaneous selections. Otherwise, only one selection will be allowed at a time.

labels (optional) The sixth argument is a pointer to an associative array relating the check box values to the user-visible labels that will be printed next to them. If not provided, the values will be used as the default.

Radio Button Group

The function radio_group() creates a set of logically related radio buttons (turning one member of the group on turns the others off).

```
print $query->radio_group(-name=>'state_name',
                          -values=>['CA','TX','AZ'],
                          -default=>'TX',
                          -linebreak=>'true',
                          -labels=>\%labels);
```

or

```
print $query->radio_group('state_name',['CA','TX','AZ'],
              'TX','true',\%labels);
print $query->radio_group(-name=>'state_name',
                          -values=>['CA','TX','AZ'],
                          -rows=2,-columns=>2);
```

When the form is processed, the selected radio button can be retrieved using:

```
$selected_radio_button = $query->param('state_name');
```

Arguments

name and values (required) The first and second arguments are the radio_button name and values, respectively. As in the pop-up menu, the second argument should be an array reference. These values are used for the user-readable labels printed next to the radio group, as well as for the values passed to your script in the query string.

default (optional) The third argument contain the values to be turned on by default.

linebreak (optional) The fourth parameter can be set to "true" to put line breaks between the buttons, creating a vertical list.

labels (optional) The sixth argument is a pointer to an associative array relating the check box values to the user-visible labels that will be printed next to them. If not provided, the values will be used as the default.

rows and columns These parameters cause radio_group() to return an HTML3-compatible table containing the radio group formatted with the specified number of rows and columns. To include row and column headings in the returned table, you can use the -rowheader and -colheader parameters. Both accept a pointer to an array of headings to use.

Submit Button

At least one submit button is necessary with every form. When the user clicks the Submit button, the farm elements are returned back to the script with the predefined GET or POST method. submit() will create the query submission button.

```
print $query->submit(-name=>'Enter',-value=>'value');
```

or

```
print $query->submit('Enter','value');
```

Arguments

name and values (optional) You need names only if there is more than one button. The name will also be used as the user-variable label. You can figure out which button was pressed by using different values for each one:

```
$which_one = $query->param('button_name');
```

Reset Button

When the user clicks the Reset button, the farm elements are restored to their values from the last time the script was called, not necessarily to the defaults.

```
print $query->reset;
```

Default Button

When the user clicks the Default button, the farm elements are restored to their default values.

```
print $query->defaults('button_label');
```

PG.PM

Our main reason for discussing Pg modules is that we have chosen PostgreSQL as the database for our embedded system design, and the Web as the media to interact with the user. The values for user interaction are taken from the database and presented to the user in the form of a Web page. The Web interface also serves the purpose of receiving user configuration that is being stored back into the database. The functional interface of the Pg module is similar to the Libpq module of the C language interface (discussed in Chapter 6, "Structured Query Language"). Next, we discuss how to incorporate the Pg module into our user interface Perl scripts.

Opening and Closing the Database

Perl's database interface module, DBI, connects Perl programs with an SQL database. The DBI is a collection of drivers for many database systems such as Oracle, Sybase, and PostgreSQL. One advantage of using DBI is the portability across several database systems, in case you decide to change the database server in the future. To include Pg as the function-oriented interface, simply add the module to the namespace and use the functions listed in the module as shown in the following code segment:

```
use DBI;
my $dbh = DBI->connect('dbi:Pg:dbname=timelog', 'nobody', '')
or die " Cannot connect to database $!\n";
```

Once you are finished with the database, the following function closes the connection to the back end and frees the connection data structure:

```
PQClear($conn);
```

or

```
$dbh->disconnect;
```

When opening a connection, a given database name is always converted to lowercase, unless it is surrounded by double quotes. All unspecified parameters are replaced by environment variables or by hard-coded defaults:

parameter	*environment variable*	*hard coded default*
host	PGHOST	localhost
port	PGPORT	5432
options	PGOPTIONS	""
tty	PGTTY	""
dbname	PGDATABASE	current userid
user	PGUSER	current userid
password	PGPASSWORD	""

Using appropriate methods, you can access almost all the fields of the returned PGconn structure.

```
$conn = Pg::setdbLogin($pghost, $pgport, $pgoptions, $pgtty, $dbname,
$login, $pwd);
```

Accessing Database Rows

The steps to fetch rows from a database table are connection, sending queries, and result. Once the connection is established, a query should be prepared.

```
$sql = "INSERT INTO employee (name, address, city, state, ph,
status,id )".
    "VALUES (\'$name\', \'$address\', \'$city\'," .
" \'$state\', \'$ph\', \'$status\', \'$id\')";
my $sth = $dbh->prepare($sql);
```

Send your queries with execute command

```
$sth->execute;
```

Check for the error if queries are not accepted.

```
if (!$dbh->commit)
{
$errstr = $dbh->errstr;
}
```

The method fetchrow can be used to sequentially fetch the next row from the server, until a NULL is returned indicating no more data.

```
While (my ($name,$address, $city, $state, $status,$ph, $id) = $sth->fetchrow)
{
        print ($name,$address, $city, $state, $status,$ph, $id);
}
```

Alternatively, you could assign the entire collection of rows into a single array.

```
@row = $result->fetchrow;
for $i ( 0 .. $#row )
{
    for $j ( 0 .. $#{$row[$i]} )
    {
            print "$row[$i][$j]\t";
    }
    print "\n";
}
```

Finally, you can close the connection.

```
$sth->finish;
$dbh->disconnect;
```

HTML Forms and Web Programming with Perl

Listing 5.1 is a Web form that requests employee information. Upon successful completion, the information is saved to a database as part of the design project of the Employee Entrance Time Log and Security System design project. We developed the first part of the project in Chapter 2 with the database schema and a simple Web form that requests login information from the user.

The first part of the script prints a form requesting time-in or time-out information, and if the user is a manager, more buttons are available to request further information. The output of the script is shown in Figure 5.1

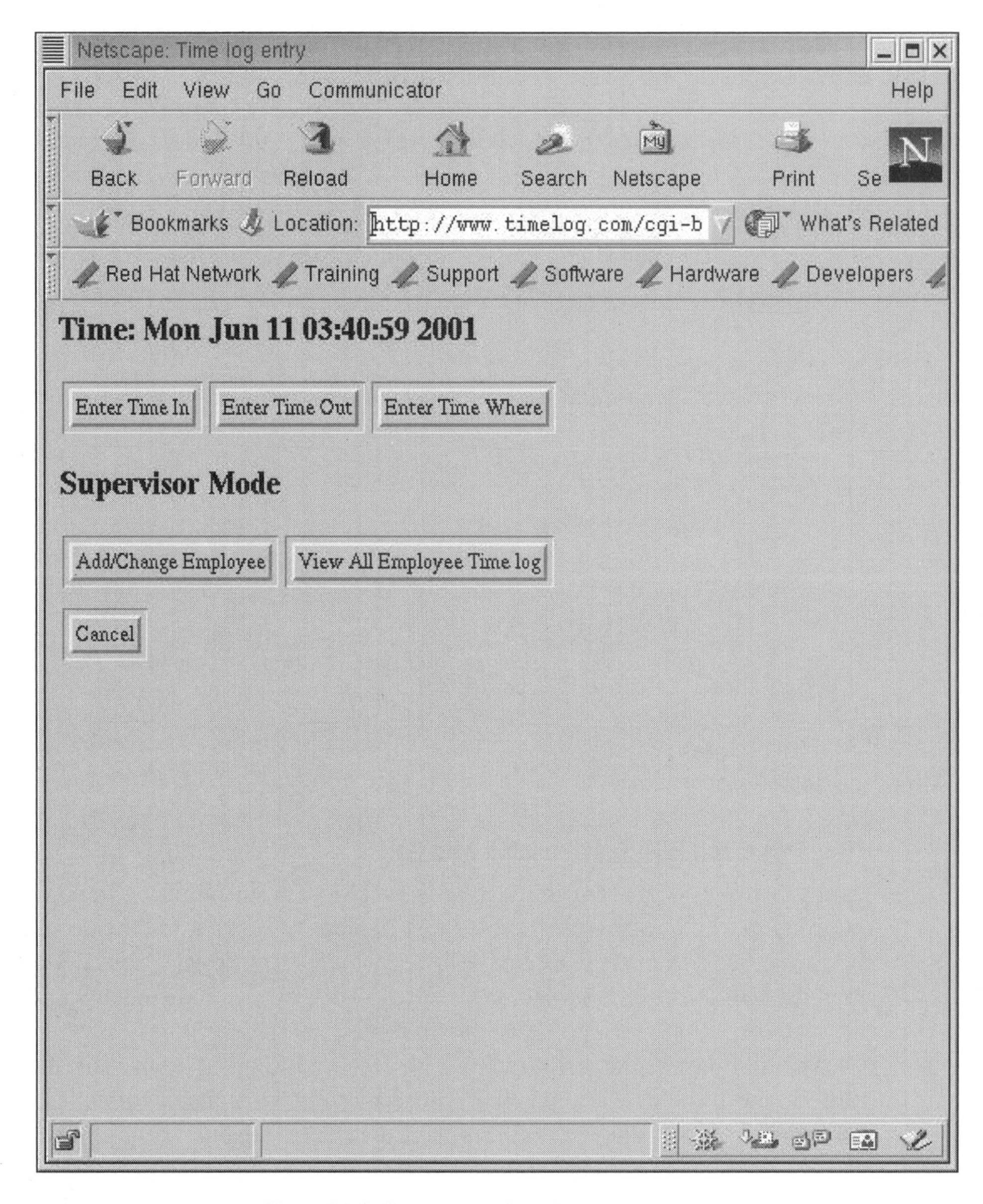

FIGURE 5.1 Forms with multiple buttons and text area.

LISTING 5.1 Verify_employee.pl file; CGI script to generate timelog form.

```
1)      #!/usr/bin/perl
2)      use CGI;
3)      use DBI;
```

```
4)      $q = new CGI;
5)      $grey = '#666666';
6)      $red  = '#FF3333';
7)      $errstr = '';
8)      $name_color = $grey;
9)      $id_color = $grey;
10)     print $q->header;
11)     print $q->start_html("Time log Action");
12)     print $q->h1(" Add a new employee ");
13)     &verify_form($q);
14)     print $q->end_html;
15)     sub verify_form()
16)     {
17)     my $complete = 1;
18)     unless ($q->param)
19)     {
20)     &print_form($q);
21)     print "<b> Nothing yet </b>";
22)     return;
23)     }
24)     if (length ($q->param('name')) lt 1)
25)     {
26)     $name_color = $red;
27)     $complete = 0;
28)     }
29)     if (length ($q->param('id')) lt 4)
30)     {
31)     $id_color = $red;
32)     $complete = 0;
33)     }
34)     if ( !$complete)
35)     {
36)     &print_form($q);
37)     print "<H2> Please fill in  the missing info </H2>";
38)     return;
39)     }
40)     &add_to_database($q);
41)     if ($errstr ne '')
42)     {
43)     print "Error: ", $errstr;
44)     return;
45)     }
46)     $q->delete_all();
47)     &print_form($q);
```

```
48)     print "<HR></BODY></HTML>";
49)     }
50)     sub print_form
51)     {
52)     my ($q) = @_;
53)     print $q->start_form;
54)     print "<STRONG STYLE=\"background: $name_color\"> NAME:</STRONG>",
            i.     $q->textfield(-name=>'name',-size=>20),$q->p,
55)     "<STRONG>Addr:</STRONG>",$q->textfield(-name=>'address',-size=>32),
            i.     $q->p,
56)     "<STRONG>City:</STRONG>",$q->textfield(-name=>'city',-size=>16),
57)     "<STRONG>State:</STRONG>",$q->textfield(-name=>'state',-size=>2),
58)     "<STRONG>ph:</STRONG>",$q->textfield(-name=>'ph',-size=>10),
            i.     $q->p,
59)     "<STRONG>Status:</STRONG>",
            i.     $q->scrolling_list(-name=>'status',
                          a.     -values=>['Employee', 'Supervisor'],
                   2. -default=>'Employee',
                          a.     -size=>1),
60)     "<STRONG STYLE=\"background: $id_color\">ID:</STRONG>",
            i.     $q->textfield(-name=>'id',-size=>4),
            ii.    $q->p;
61)     print $q->submit;
62)     print $q->reset;
63)     print $q->endform;
64)     }
65)     sub add_to_database()
66)     {
67)     my ($q) = @_;
68)     my $name = $q->param('name');
69)     my $id = $q->param('id');
70)     my $address = $q->param('address');
71)     my $city = $q->param('city');
72)     my $state = $q->param('state');
73)     my $status = $q->param('status');
74)     my $ph = $q->param('state');
75)     my $dbh = DBI->connect('dbi:Pg:dbname=timelog', 'nobody', '')
            or die " Cannot connect to database $!\n";
76)     my $sql = "INSERT INTO employee (name, address, city, state, ph, status,
id ) " .
77)     "VALUES (\'$name\', \'$address\', \'$city\'," .
78)     " \'$state\', \'$ph\', \'$status\', \'$id\')";
79)     my $sth = $dbh->prepare($sql);
80)     $sth->execute;
```

```
81)     if (!$dbh->commit)
82)     {
83)     $errstr = $dbh->errstr;
84)     }
85)     #my ($name,$address, $city, $state, $status,$ph, $id) = $sth->fetchrow;
86)     #$sth->finish;
87)     $dbh->disconnect;
88)     }
```

BUILT-IN FUNCTIONS IN PERL

The following is a list of functions built in to the Perl interpreter. The functions can be used as statements in a Perl program.

-X	run a file test
abs	absolute value function
accept	accept an incoming socket connect
alarm	schedule a SIGALRM
atan2	arctangent of Y/X
bind	binds an address to a socket
binmode	prepare binary files on old systems
bless	create an object
caller	get context of the current subroutine call
chdir	change your current working directory
chmod	changes the permissions on a list of files
chomp	remove a trailing record separator from a string
chop	remove the last character from a string
chown	change the owership on a list of files
chr	get character this number represents
chroot	make directory new root for path lookups
close	close file (or pipe or socket) handle
closedir	close directory handle
connect	connect to a remove socket
continue	optional trailing block in a while or foreach
cos	cosine function

crypt	one-way passwd-style encryption
dbmclose	breaks binding on a tied dbm file
dbmopen	create binding on a tied dbm file
defined	test whether a value, variable, or function is defined
delete	deletes a value from a hash
die	raise an exception or bail out
do	turn a BLOCK into a TERM
dump	create an immediate core dump
each	retrieve the next key/value pair from a hash
endgrent	be done using group file
endhostent	be done using hosts file
endnetent	be done using networks file
endprotoent	be done using protocols file
endpwent	be done using passwd file
endservent	be done using services file
eof	test a filehandle for its end
eval	catch exceptions or compile code
exec	abandon this program to run another
exists	test whether a hash key is present
exit	terminate this program
exp	raise e to a power
fcntl	file control system all
fileno	return file descriptor from filehandle
flock	lock an entire file with an advisory lock
fork	create a new process just like this one
format	declare a picture format with use by the write() function
formline	internal function used for formats
getc	get the next character from the filehandle
getgrent	get next group record
getgrgid	get group record given group user ID
getgrnam	get group record given group name
gethostbyaddr	get host record given its address

gethostbyname	get host record given name
gethostent	get next hosts record
getlogin	return who logged in at this tty
getnetbyaddr	get network record given its address
getnetbyname	get networks record given name
getnetent	get next networks record
getpeername	find the other hend of a socket connection
getpgrp	get process group
getppid	get parent process ID
getpriority	get current nice value
getprotobyname	get protocol record given name
getprotobynumber	get protocol record numeric protocol
getprotoent	get next protocols record
getpwent	get next passwd record
getpwnam	get passwd record given user login name
getpwuid	get passwd record given user ID
getservbyname	get services record given its name
getservbyport	get services record given numeric port
getservent	get next services record
getsockname	retrieve the sockaddr for a given socket
getsockopt	get socket options on a given socket
glob	expand filenames using wildcards
gmtime	convert UNIX time into record or string using Greenwich time
goto	create spaghetti code
grep	locate elements in a list test true against a given criterion
hex	convert a string to a hexadecimal number
import	patch a module's namespace into your own
int	get the integer portion of a number
ioctl	system-dependent device control system call
join	join a list into a string using a separator
keys	retrieve list of indices from a hash
kill	send a signal to a process or process group

last	exit a block prematurely
lc	return lower-case version of a string
lcfirst	return a string with just the next letter in lower case
length	return the number of bytes in a string
link	create a hard link in the filesytem
listen	register your socket as a server
local	create a temporary value for a global variable (dynamic scoping)
localtime	convert UNIX time into record or string using local time
log	retrieve the natural logarithm for a number
lstat	stat a symbolic link
m//	match a string with a regular expression pattern
map	apply a change to a list to get back a new list with the changes
mkdir	create a directory
msgctl	SysV IPC message control operations
msgget	get SysV IPC message queue
msgrcv	receive a SysV IPC message from a message queue
msgsnd	send a SysV IPC message to a message queue
my	declare and assign a local variable (lexical scoping)
next	iterate a block prematurely
no	unimport some module symbols or semantics at compile time
oct	convert a string to an octal number
open	open a file, pipe, or descriptor
opendir	open a directory
ord	find a character's numeric representation
pack	convert a list into a binary representation
package	declare a separate global namespace
pipe	open a pair of connected filehandles
pop	remove the last element from an array and return it
pos	find or set the offset for the last/next m//g search
print	output a list to a filehandle
printf	output a formatted list to a filehandle
prototype	get the prototype (if any) of a subroutine

push	append one or more elements to an array
q/STRING/	singly quote a string
qq/STRING/	doubly quote a string
quotemeta	quote regular expression magic characters
qw/STRING/	quote a list of words
qx/STRING/	backquote quote a string
rand	retrieve the next pseudorandom number
read	fixed-length buffered input from a filehandle
readdir	get a directory from a directory handle
readlink	determine where a symbolic link is pointing
recv	receive a message over a socket
redo	start this loop iteration over again
ref	find out the type of thing being referenced
rename	change a filename
require	load in external functions from a library at runtime
reset	clear all variables of a given name
return	get out of a function early
reverse	flip a string or a list
rewinddir	reset directory handle
rindex	right-to-left substring search
rmdir	remove a directory
s///	replace a pattern with a string
scalar	force a scalar context
seek	reposition file pointer for random-access I/O
seekdir	reposition directory pointer
select	reset default output or do I/O multiplexing
semctl	SysV semaphore control operations
semget	get set of SysV semaphores
semop	SysV semaphore operations
send	send a message over a socket
setgrent	prepare group file for use
sethostent	prepare hosts file for use

setnetent	prepare networks file for use
setpgrp	set the process group of a process
setpriority	set a process's nice value
setprotoent	prepare protocols file for use
setpwent	prepare passwd file for use
setservent	prepare services file for use
setsockopt	set some socket options
shift	remove the first element of an array, and return it
shmctl	SysV shared memory operations
shmget	get SysV shared memory segment identifier
shmread	read SysV shared memory
shmwrite	write SysV shared memory
shutdown	close down just half of a socket connection
sin	return the sin of a number
sleep	block for some number of seconds
socket	create a socket
socketpair	create a pair of sockets
sort	sort a list of values
splice	add or remove elements anywhere in an array
split	split up a string using a regexp delimiter
sprintf	formatted print into a string
sqrt	square root function
srand	seed the random number generator
stat	get a file's status information
study	optimize input data for repeated searches
sub	declare a subroutine, possibly anonymously
substr	get or alter a portion of a stirng
symlink	create a symbolic link to a file
syscall	execute an arbitrary system call
sysread	fixed-length unbuffered input from a filehandle
system	run a separate program
syswrite	fixed-length unbuffered output to a filehandle

tell	get current seekpointer on a filehandle
telldir	get current seekpointer on a directory handle
tie	bind a variable to an object class
time	return number of seconds since 1970
times	return elapsed time for self and child processes
tr///	transliterate a string
truncate	shorten a file
uc	return upper-case version of a string
ucfirst	return a string with just the next letter in upper case
umask	set file creation mode mask
undef	remove a variable or function definition
unlink	remove one link to a file
unpack	convert binary structure into normal perl variables
unshift	prepend more elements to the beginning of a list
untie	break a tie binding to a variable
use	load in a module at compile time
utime	set a file's last access and modify times
values	return a list of the values in a hash
vec	test or set particular bits in a string
wait	wait for any child process to die
waitpid	wait for a particular child process to die
wantarray	get list vs array context of current subroutine call
warn	print debugging info
write	print a picture record
y///	transliterate a string

SUMMARY

As we have shown, Perl as a language is ideally suited for designing the user interface of an embedded system. We covered the Perl syntax from a programming point of view, and described the CGI and the Pg modules as an object-oriented interface. The Perl program structure is very simple, and the absence of strict type checking gives a programmer a great advantage over the C++ object-oriented interface. The search and replacement capability as a built-in option in Perl simplifies a lot of Web-oriented application design. Perl is gaining respect in the programming community, and more and more modules are being introduced covering a wide variety of programming needs.

6 Structured Query Language

In this chapter

- Relational Database
- SQL Statements
- SQL Queries
- Joining Tables
- Data Types Supported by PostgreSQL
- Application Programming Using C

INTRODUCTION

Let us review the design philosophy of our embedded system design one more time: An embedded system includes a device driver to access the hardware, a client server mechanism to communicate with the device driver, and saves the data in a database. A user interface, on the other hand, picks up the data from the database, creates a Web page out of it with the help of a Perl or C script, and passes it along to a Web server that communicates with the user and presents the page on demand. The process is reversed when a user decides to change some parameters: the user fills out a Web form, the form is returned to the server, and the database is updated. The system interface picks up the new parameter from the database and makes the appropriate system changes.

As you can see, the database plays a central role of communication between the system and the user interface. Interaction with the database is incomplete without an understanding of the Structured Query Language, or SQL. SQL is the native language of most database servers. We discuss the alternative for low-end embedded systems later in the chapter. A database is a catalogue of information presented in a tabulated form, where each table has a set of columns, forming rows of information on related items. We need a database for persistence. We need to know where the information is, even when the power is down. A good schema of the database tables is also a good source of documentation on your embedded system components. You can trace the flow of information by simply monitoring the changes in the database at different points of program execution. Granted, a full-blown database SQL server is not in the best interest of many low-end embedded systems, but sooner or later, an embedded system designer will encounter a large database that requires a SQL server. Therefore, learn SQL queries; they are simple, intuitive, and best of all, the pay scale for a database programmer is much higher than that for an embedded system designer—so, there is light at the end of the tunnel.

RELATIONAL DATABASE

You might have several tables in the same database with different sets of information that are somehow related. For example, a database of company employees might have a table containing rows of information for each employee—employee number, hiring date, hourly wages, and so forth. Another table might show employee attendance with columns for time-in and time-out information, and a column for employee number to identify to whom the row of information belongs. The fact that the employee number column is the same for both tables creates a relationship between the two tables. A relational database knows how to gather information from different tables just by knowing the relationship between the tables. You will see an example later in the chapter when we develop SQL queries for our embedded system design of an employee entrance log.

Database Schema

A database schema is simply a description of all the tables and the associated columns that make up the table. The schema should identify the key columns in each table that form the relationship among them, and the index column if there is any. The index column could help itemize the rows in the table, but it is not mandatory, whereas the key columns are necessary if you want to establish a relationship among different tables. Let us create a schema of the relational database for our embedded system design project "Employee Entrance Time Log." This is a continuation of the project we discussed in Chapter 2, "The User Interface." The Web page in that chapter was developed to show a simple user interface for entering employee information and viewing the time log of each employee. Now, we go one step further and establish the database and its queries to get information from the database.

We will call our database "Payroll," and have two tables, "EmployeeInfo" and "tm_log" (Table 6.1).

TABLE 6.1 Database Payroll Schema

Table Employee		
Column 1	ID	(Char 6)—- > Primary Key
Column 2	LastName	(Char 16)
Column 3	FirstName	(Char 16)
Column 4	MI	(Char 2)
Column 5	Active	(Char 2)
Column 6	Supervisor	(Char 6)
Column 7	Payrate	(Char 6)
Table TimeLog		
Column 1	ID	(Char 6) —- > Foreign Key
Column 2	TimeIn	(TIMESTAMP)
Column 3	TimeOut	(TIMESTAMP)

The EmployeeInfo table is the employee information table, and the TimeLog table is the attendance information table. The two tables are kept separate, and have only the employee number column in common. It is possible to have everything in one table, but we would end up having redundant information with one long list of rows with repeated information of employee number, hiring date, and hourly wages, together with the time-in and time-out information.

We avoid redundancy by having two separate tables with a common key column only, which makes it a relational database. Next, a tutorial on SQL queries, but first, some basic nomenclatures in SQL.

Keys

A *primary key* in database tables is a column or set of columns that uniquely identifies the rest of the data in any given row. For example, in the Employee table, the ID column being a primary key uniquely identifies that row. Having a primary key means that no two rows can have the same ID. This is how we differentiate employees with the same last name or first name, or any other column value that might be common among employees.

A *foreign key* is a column in a table where that column is a primary key of another table. What this means is that for each data in a foreign key column, there must exist a corresponding data in the other table where that column is the primary key. In database terminology, this correspondence is known as *referential integrity*. For example, in the Timelog table, the ID column is the foreign key to the primary key of the Employee table. In order to have an entry in the Timelog table, the ID should belong to an employee who has an entry in the Employee table.

Join

Multiple tables are created in a database to avoid unnecessary repetition of column values. However, eventually, we have to combine the information from different tables to produce a desired outcome. For example, what if we want to find out the names of all employees who were present at certain time? Obviously, the timelog information from the TimeLog table is joined with the name information from the Employee with the same ID. Such an operation of combining tables is called JOIN, which relates the data in two or more tables. You will see more examples later in the chapter.

SQL STATEMENTS

A database server such as PostgreSQL is essentially a database management system. To communicate with the server, a request is submitted via a backend. The interactive backend for PostgreSQL is *psql*, and the request statements submitted to extract information are called SQL *queries*. Please revisit Chapter 2 to see how PostgrSQL was installed and how the database tables were created using *psql* commands. The tutorial in Chapter 2 showed some SQL queries in action, to create the database tables and add entries to them. The best help is already provided by the *psql* itself, and you can glance the basic categories by simply entering the help command "\h" and "\?" at the psql command prompt. In this chapter, we will discuss how the queries work and various data types that are being accepted. We will create some of our own queries that we will use in our own embedded system design project "Employee Entrance and Time Log." One good thing about SQL queries is that you can test them interactively using the *psql* client. The same queries can be presented programmatically via Perl/CGI scripts as well as executable C programs. Those are presented in Chapter 9, "Embedded System Design Projects."

The statements that comprise extracting information from the database are called *queries* and are being implemented as *SELECT*statements. The process of joining two or more tables dynamically for the purpose of extracting common information is accomplished through *JOIN* statements as explained in the previous section. A *natural join* is when a foreign key of one table is joined with its parent key in the other table. With a natural join of the primary key of the employee table with the foreign key of tm_log table, we can find all the timestamps of a particular employee in our *payroll* database.

SQL statements can be divided into the following categories:

- Creating, changing, and dropping tables
- Entering, deleting, and changing data
- Setting constraints on data values
- Relational integrity using keys, and joining tables
- Review information from the tables
- Using expressions in statements

Creating, Changing, and Dropping Tables

The tables in a database are comprised of columns and rows. You can add, delete, or change the composition of tables interactively using a backend client such as "psql."

In order to do the examples presented in this section, you need to log in as user "postgres" and create a database "timelog" using createdb command as follows:

```
#su postgres
$createdb timelog
```

Execute the interactive program "psql."

```
$psql timelog
```

ON THE CD

The file in the companion CD directory /book/chapter6/employee_schema can be loaded and executed using the following psql command:

```
timelog=# \i /book/chapter6/employee_schema
```

The CREATE TABLE command creates a new table. It is initially empty, meaning there are no rows. The format for CREATE TABLE is:

```
CREATE TABLE tablename ({columnname datatype[(size)}.,..);
```

For example:

a) The following statement creates the table employee as defined in the schema in Table 6.1:

```
timelog=# create table employee (ID char (10), password char (10),
LastName char(16), FirstName char(16), MI char(2), Active (2),
Supervisor  char(10));
```

b) The following statement creates the table entrancelog as defined in the schema in Table 6.1:

```
timelog=# create table entrancelog (ID char (10), timein timestamp,
timeout timestamp);
```

The ALTER TABLE command changes the table parameter after it is created. The following is the format for ALTER TABLE commands:

```
ALTER TABLE tablename
    { ADD [COLUMN] column  [*] type }
    | { ALTER [COLUMN] column name
           { SET DEFAULT default option } | { DROP DEFAULT}}
    | {  RENAME column TO new column name}
    | { ADD table constraints definition }
    | {  RENAME TO new table name };
```

We can perform the following actions with the ALTER TABLE command.

Add a Column to a Table

The following statement adds a new column of *payrate* to the employee table:

```
timelog=#  Alter  table employee  add payrate  float8;
```

Alter Table Column Set Default

We can add default value to column type.

```
timelog=#  Alter  table employee Alter payrate Set default 0.0;
```

Alter Table Column Drop Default

The following statement drops the default value assigned to a column.

```
timelog=#  Alter  table employee Alter payrate DROP default;
```

Rename Column

The following statement renames the column payrate to pay.

```
timelog=#  Alter  table employee RENAME payrate TO pay;
```

Rename Table

The following statement renames the table employee to personel.

```
timelog=#  Alter  table employee RENAME TO personel;
```

The DROP TABLE command removes the table from the database, but the table must be emptied before it can be dropped.

Format

DROP TABLE tablename;

Entering, Deleting, and Changing Data

Once a table is defined, you can insert, delete, and change column values using psql commands.

Entering Values into a Table

The INSERT INTO command allows you to enter rows into tables. The database manager will check for all the constraints that have been placed before, as well as add default values to columns whose value are not specified. If a default value is not specified at table creation time, a NULL value will be specified for the column.

Format

INSERT INTO table name VALUES (value ., ..);

The order of parameters followed by the keyword VALUES should match the corresponding column in the table. For example, the following statement creates one row of information into the employee table, initializing ID= '1234', First-Name= 'John', LastName= 'Public', MI= 'Q', active= 'A', Supervisor= '1234', payrate= '20.00':

```
timelog=#  INSERT INTO employee VALUES ('1234',  'Public', 'John',
'Q', 'A', '1234', '20.00');
```

Naming Columns for Insert

You do not have to insert values for all the columns at once. You can define the name of the column for which the values are being specified. The following query will only initialize the ID, LastName, and payrate columns:

```
timelog=#  INSERT INTO employee (ID, LastName, payrate) VALUES ('1234',
'Public','20.00');
```

Defining NULLs

If values are not known beforehand, you can specify NULL in the value position for the column:

```
timelog=#  INSERT INTO employee VALUES ('1234',
'Public',NULL,NULL,NULL,'20.00');
```

Using Query to Insert Values

You can essentially duplicate the contents of rows by specifying the result of another query. The following example creates a new table *supervisor1234* that contains all the rows found in the table *employee* that match the supervisor ID "1234."

```
timelog=#  create table supervisor1234 (ID char (10), password char
(10), LastName char(16), FirstName char(16), MI char(2), Active (2),
Supervisor  char(10));

timelog=#  INSERT INTO supervisor1234 SELECT * FROM employee where ID =
'1234';
```

Deleting Rows from a Table

The DELETE FROM statement removes rows from a table. The following statement deletes all rows from the employee table:

```
timelog=#  DELETE * FROM Employee;
```

You can specify a condition using the WHERE clause that tests for a true condition before a row is deleted. The following statement deletes only those rows that contain Active value not equal to 'A'.

```
timelog=#  DELETE FROM employee WHERE Active <> 'A';
```

The following statement deletes the row for employee ID 1234:

```
timelog=#  DELETE FROM employee WHERE ID = '1234';
```

Changing Column Values

The UPDATE command modifies the values in a row. You can use the WHERE clause to specify a condition that affects a specific row that matches the condition. The following statement changes the supervisor ID to 3456 for all the rows with the Supervisor ID 1234:

```
timelog=#  UPDATE employee SET Supervisor = '3456' WHERE Supervisor =
'1234';
```

Inserting NULL Values

The UPDATE command can be used to insert a NULL value just like the INSERT command by simply not specifying a value:

```
timelog=#  UPDATE employee;
```

Setting Constraints

Constraints can be placed on data values as well as on column types. The data value constraints can be a range of values, whereas the system constraints are criteria being applied to columns, such as excluding NULLS, specifying primary keys, foreign keys, primary keys on more than one column, and UNIQUE. You can apply constraints at the time of table creation or later using the ALTER command:

```
CREATE TABLE tablename ({columnname datatype [(size)]}.,..,
[table constraint (columnname)]};
```

You can also specify constraints when individual columns are specified:

```
CREATE TABLE tablename ({columnname datatype column constraints
[(size)]}.,..);
```

Primary Keys

Specifying primary keys to a column or group of columns sets aside criteria for joining two or more tables, but it is actually a constraint. The column by default cannot accept NULL values, and the values are UNIQUE. The following statement specifies that FirstName and LastName together form the primary key:

```
timelog=#  create table employee (ID char (10), password char (10),
LastName char(16), FirstName char(16), MI char(2), Active (2),
Supervisor  char(10) PRIMARY KEY (LastName, FirstName));
```

You can also specify a column to be a constraint with a primary key as shown in the following statement, creating the ID column to be the primary key of the employee table:

```
timelog=#  create table employee (ID char (10)PRIMARY KEY , password
char (10), LastName char(16), FirstName char(16), MI char(2), Active
(2), Supervisor  char(10) );
```

The Foreign Key as a Referential Integrity

The foreign key in a table column creates a reference to its parent column in another table. The column ID in the employee table is being referenced in another table tm_log column ID as the foreign key in that table, as shown in the following statement:

```
timelog=#  create table tm_log (ID char (10), password char (10),
LastName char(16), FirstName char(16), MI char(2), Active (2),
Supervisor  char(10)  FOREIGN KEY (ID) REFERENCES employee (ID) );
```

Another way to specify a foreign key is through constraints on columns instead of tables:

```
timelog=#  create table tm_log (ID char (10), FOREIGN KEY (ID)
REFERENCES employee (ID)   , password char (10), LastName char(16),
FirstName char(16), MI char(2), Active (2), Supervisor  char(10)  );
```

If the two tables have the same column name for primary and foreign keys, the name can be omitted when the foreign is being referenced, as shown in the following example of creating tm_log:

```
timelog=#  create table tm_log (ID char (10), REFERENCES employee ,
password char (10), LastName char(16), FirstName char(16), MI char(2),
Activechar  (2), Supervisor  char(10) );
```

Using a Foreign Key to Check Against a Known Set of Values

If you want to restrict the input values to be a prescribed range, place the range in a table and reference it through a foreign key. For example, if we want the values in the column Active in the table Employee to have values "A" and "P," we create a table ActiveState as shown here and reference it in the employee table as a foreign key:

```
create table ActiveState(Active char (2) PRIMARY KEY);
    insert into ActiveState Values ('A');
```

```
    insert into ActiveState Values ('P');
create table employee (ID char (10), password char (10), LastName
char(16), FirstName char(16), MI char(2), Active (2)REFERENCE
ActiveState, Supervisor  char(10) PRIMARY KEY (LastName, FirstName));
```

Constraints on Column Values Using CHECK

This type of constraint prevents quite a few data entry problems such as entering a negative number, value exceeding a limit, and so forth. The following example demonstrates restricting employee ID to have values between "0001" and "9999":

```
timelog=#  create table employee (ID char (10)CHECK (ID <"9999" AND ID >
"0000"), password char (10), LastName char(16), FirstName char(16), MI
char(2), Active (2), Supervisor  char(10));
```

Constraints on Multiple Column Values with CHECK

You can also specify constraints on multiple columns simultaneously. The following SQL statement restricts the employee ID to values between "0001" and "9999," and the Active column to have value either "A" or "P":

```
timelog=#  create table employee (ID char (10), password char (10),
LastName char(16), FirstName char(16), MI char(2), Active (2),
Supervisor  char(10) CHECK ((ID <"9999" AND ID > "0000") AND ("ACTIVE =
"A" OR ACTIVE="P")) );
```

Uniqueness of Values Using UNIQUE

The Primary key column is usually required to be unique, but you can also specify a column to have UNIQUE value that is different in each row. There is a difference between the uniqueness of primary key and a column with unique constraint:

- The primary key is restricted to one column or group of columns, whereas the UNIQUE constraint can be placed on any number of columns.
- NULLs are not allowed on primary columns, but UNIQUE columns can have a NULL value.

The following statement places the UNIQUE constraint on the FirstName and LastName columns. Together, the column values have to be unique.

```
timelog=#  create table employee (ID char (10), password char (10),
LastName char(16), FirstName char(16), MI char(2), Active (2),
Supervisor  char(10) UNIQUE (LastName,FirstName ));
```

Excluding NULL Values from Columns

NULL essentially means having no column value for the row being specified. You can avoid having NULL values with the constraint NOT NULL as shown in the following statement, which forces the FirstName to have a value:

```
timelog=#  create table employee (ID char (10), password char (10),
LastName char(16), FirstName char(16) NOT NULL, MI char(2), Active (2),
Supervisor  char(10) UNIQUE (LastName,FirstName ));
```

DEFAULT Values Assignment

You saw the DEFAULT statement in the description of the ALTER TABLE command. You can specify the DEFAULT values when the table is created. The following example sets aside a default value of 'A' for the column Active in the employee table:

```
timelog=#  create table employee (ID char (10), password char (10),
LastName char(16), FirstName char(16) NOT NULL, MI char(2), Active (2)
DEFAULT = 'A' , Supervisor  char(10) UNIQUE (LastName,FirstName ));
```

Transaction and Locks

Changing values in a row simultaneously by two different users is prevented by the mechanism of Transaction built into the postgresql database server. It is like having a read, modify, write operation atomically without being disturbed by other sources. Each statement is executed as a transaction, but you can apply a lock into a group of statements using BEGIN WORK and COMMIT WORK statements. For example:

```
timelog=#  BEGIN WORK;
timelog=#  UPDATE employee SET Firstname = 'David' where id = '1111';
timelog=#  COMMIT WORK;
```

Foreign Key and Primary Key from the Same Table

In order to explain a situation in which a foreign key refers back to the primary key from the same table, we will add a column name *Supervisor* to our employee table. Obviously, every employee has a supervisor except for the top-level manager. The unique aspect of this situation is that the foreign key must allow NULL as column value.

```
create table employee (ID char (10) PRIMARY KEY, password char (10),
LastName char(16), FirstName char(16) NOT NULL, MI char(2), Active (2)
DEFAULT = 'A' , Supervisor  char(10) REFERENCE Employee);
```

Miscellaneous SQL Statements

The SQL supports special functions called *aggregates* that operate on the entire group of columns and produce a single value result.

Aggregate Functions

SUM, AVG, MAX, MIN, and COUNT are called *aggregate functions*, as they return a single result after performing the desired operation on the column values.

- **SUM** () gives the total of all the rows for a numeric column.

```
SELECT SUM(feet) FROM Depth;
```

- **AVG** () gives the average of the given column.

```
SELECT AVG(feet) FROM Depth;
```

- **MAX** () gives the largest figure in the given column.

```
SELECT MAX(feet) FROM Depth;
```

- **MIN** () gives the smallest figure in the given column.

```
SELECT MIN(feet) FROM Depth;
```

- **COUNT**(*) gives the number of rows satisfying the conditions.

```
SELECT COUNT(feet) FROM Depth;
```

Let us say that we have computed the salary of employees using the hourly_wages and the time clocked in by the employees, and created a new table *Salary* that contains employeeID and wages earned. We can compute the sum, average, maximum, minimum, and count with the following query:

```
SELECT SUM(wages), AVG(wages), MAX (wages), MIN (wages), Count (ID)
FROM Salary;
```

This query shows the total of all salaries in the table, and the average, maximum, and minimum salary of all the entries in the table.

SQL QUERIES

The process of extracting information from database is done via the keyword SELECT.The statements using SELECT are called *querystatements*. The following is a simple statement that extracts information from all the rows from the employee table employee:

```
timelog=#  SELECT * FROM Employee;
```

Selecting Certain Columns Only

If you only need information from the LastName and FirstName from the Employee table, set the query as follows:

```
timelog=#  SELECT LastName, FirstName FROM Employee;
```

Using DISTINCT to Eliminate Redundant Data

If a table has several values repeated, you can ask a query to eliminate the redundancy by adding DISTINCT to the query statement:

```
timelog=#  SELECT DISTINCT ID FROM tm_log;
```

Select ALL Rows

As opposed to DISTINCT, * extracts information from every row:

```
timelog=#  SELECT * From tm_log;
```

Qualifying Queries with the WHERE Clause

You can narrow your extracted list using the WHERE clause. For example, the following query will extract all the time-in and time-out information from the tm_log table for the employee ID 1234:

```
timelog=#  SELECT time_in time_out From tm_log where ID = '1234';
```

Arithmetic Comparison Operators in the WHERE Clause

You can build complex queries with arithmetic comparison operators with the WHERE clause. SQL recognizes the following operators:

=	Equal to
>	Greater than

<	Less than
>=	Greater than or equal to
<=	Less than or equal to
<>	Not equal to

Boolean Operators in the WHERE Clause

The SQL statements in the WHERE clause are executed only if the expression evaluates to a true condition, as you saw in the arithmetic comparison operators in the previous example. You can modify the result of the expression with the following logical operators:

NOT	Tests a single operand and the Boolean result is inverted; the true becomes false, and vice versa.
AND	Takes two operands, and the result is true when both operands are true.
OR	Takes two operands, and the result is true when any one operand is true.

The following expression returns all the time_in values for the employee ID '1234' between January and February:

```
timelog=#  SELECT time_in  From tm_log where (( ID = '1234' ) AND (
time_in > 01/01/01 AND time_in < 02/01/01));
```

Boolean Operator UNKNOWN

SQL also supports a third-level logical operator, UNKNOWN, which is similar to false when evaluated in an arithmetic expression. However, the logical operator NOT UNKNOWN does not change the value of UNKNOWN; it still remains UNKNOWN. The following truth table clarifies the result of the three-level logic supported by SQL:

Truth Table Boolean Operator NOT

NOT	TRUE	false
NOT	FALSE	true
NOT	UNKNOWN	unknown

Truth Table Boolean Operator AND

	TRUE	***FALSE***	***UNKNOWN***
TRUE	true	false	UNKNOWN
FALSE	false	false	false
UNKNOWN	unknown	false	unknown

Truth Table Boolean Operator OR

	TRUE	***FALSE***	***UNKNOWN***
TRUE	true	true	true
FALSE	true	false	false
UNKNOWN	unknown	false	unknown

The IN Operator

You can provide a list of values in the WHERE clause instead of several OR statements. The IN operator is an abbreviated form of OR. The following statement extracts time_in information for the ID 1234 and 3456:

```
SELECT time_in  From tm_log where ID IN ('1234', '3456' );
```

The BETWEEN Operator

The BETWEEN operator is similar to IN, except a range of values are computed and tested against. The effect is the same as "<=" and ">=". The following statement extracts time_in information for all employee IDs between 1234 and 3456 with alphabetical ordering:

```
SELECT time_in  From tm_log where ID BETWEEN '1234 AND '3456';
```

The LIKE Operator

LIKE is ideally suited for string search. The given string is searched in column values, and a match in any part returns a true value. You can provide the following special characters in the search string that alter the search criteria:

- The "_" underscore will match any character; for example, "_at" will match "hat" and "cat."
- The "%" percent character will match 0 or more sequences of characters in the string; for example, "%at" will match "at," "cat," and "brat."
- You can override the special character interpretation by preceding it with an ESCAPE character; for example, '/%' ESCAPE '/', will look for the character % in the string.

The following statement extracts time_in information for all employee IDs with string '21' in any part of the value:

```
SELECT time_in  From tm_log where ID LIKE '21' ;
```

The IS NULL Operator

A NULL in a column value indicates that no value has been specified, and this is where the third logic level UNKNOWN comes in handy. It avoids ambiguity when you search for false conditions where no value has been entered. The following statement prints all the employees whose MI has not been entered:

```
SELECT *  From employee where MI IS NULL;
```

JOINING TABLES

The most important feature of the relational database is the ability to combine tables based on a given relationship. We already discussed the primary key and foreign key concepts earlier in the chapter, and the criteria for establishing links between tables. Now, we discuss queries based on joining multiple tables and different ways of combining the WHERE clause to extract the desired information that is scattered in different tables. Suppose we want to find out the names of all employees who were present on a specific day and worked for supervisor ID 1234, given the sample database payroll with the table *employee* and *tm_log*. Obviously, the names have to come from the Employee table, and time-in information has to come from the table tm_log. We need a query that will combine the two tables and then extract information in a manner that satisfies the search criteria. Such an operation is called *join*. When naming columns from different tables that have similar names, in the same query statement, we need to fully qualify the column names by prepending the table name followed by a dot followed by the column name. The following query results in extracting all employees who were present on Jan 1, 2001 and work for supervisor ID 1234:

```
SELECT employee.FirstName FROM employee, tm_log
WHERE (tm_log.ID = employee.ID) AND (time_in = '01/01/01') AND
(Employee.Supervisor = '1234');
```

Notice that the dot notations were omitted when there was no ambiguity in expressing the column names.

Using Subqueries

One way to extend your search criteria is to embed a subquery in the WHERE clause. The results from the inner queries are tested by the outer query to determine a TRUE condition, and based on that, the SQL statements are executed. Suppose you wanted to find out the timestamps of an employee whose last name is "public." You can create the following query that has a subquery buried inside the main query:

```
timelog=#  SELECT tm_log.time_in FROM tm_log
WHERE tm_log.ID = (SELECT employee.ID FROM employee
WHERE employee.LastName = 'public');
```

DATA TYPES SUPPORTED BY POSTGRESQL

The following is a list of the data types supported by PostgreSQL. You can get a complete list by entering the command option '\dT' on the psql command prompt.

SET	Set of tuples
abstime	Absolute, limited-range date and time (Unix system time)
aclitem	Access control list
bit	Fixed-length bit string
bool	Boolean, "true"/"false"
box	Geometric box "(lower left, upper right)"
bpchar	Char(length), blank-padded string, fixed storage length
bytea	Variable-length string, binary values escaped
char	Single character
cid	Command identifier type, sequence in transaction ID
cidr	Network IP address/netmask, network address
circle	Geometric circle "(center,radius)"

date	ANSI SQL date
filename	Filename used in system tables
float4	Single-precision floating-point number, 4-byte storage
float8	Double-precision floating-point number, 8-byte storage
inet	IP address/netmask, host address, netmask optional
int2	–32,000 to 32,000, 2-byte storage
int2vector	Array of 16 int2 integers, used in system tables
int4	–2 billion to 2 billion integer, 4-byte storage
int8	~18-digit integer, 8-byte storage
interval	@ <number> <units>, time interval
line	Geometric line "(pt1,pt2)"
lseg	Geometric line segment "(pt1,pt2)"
lztext	Variable-length string, stored compressed
macaddr	XX:XX:XX:XX:XX, MAC address
money	$d,ddd.cc, money
name	31-character type for storing system identifiers
numeric	Numeric(precision, decimal), arbitrary precision number
oid	Object identifier(oid), maximum 4 billion
oidvector	Array of 16 oids, used in system tables
path	Geometric path "(pt1,...)"
point	Geometric point "(x, y)"
polygon	Geometric polygon "(pt1,...)"
regproc	Registered procedure
reltime	Relative, limited-range time interval (Unix delta time)
smgr	Storage manager
text	Variable-length string, no limit specified
tid	(Block, offset), physical location of tuple
time	hh:mm:ss, ANSI SQL time
timestamp	Date and time
timetz	hh:mm:ss, ANSI SQL time
tinterval	(abstime,abstime), time interval
unknown	The value will be defined later

varbit	Fixed-length bit string
varchar	Varchar(length), nonblank-padded string, variable storage length
xid	Transaction ID

APPLICATION PROGRAMMING USING C

ON THE CD

The PostgreSQL database management system as a server needs a client system to communicate with. The C language interface for the client setup is provided by the libpq library in the /usr/lib directory, and the exported function prototypes are provided in the header file libpq-fe.h in the /usr/include/pgsql directory. Listing 6.1 is a simple C language application that performs a dump of database contents, and inserts a new value into the database. The program demonstrates the use of SQL statements to communicate with the database, assuming you already created the database using the psql interpreter.

- Lines 1 through 11 are the include files for prototype.
- Lines 12 through 13 are the Pgconn and Pgresult structure as defined in the file libpq-fe.h.
- Lines 17 through 33 are a saveData routine that takes an input string ID.
- Line 20 opens the databases.
- Line 21 checks for a bad connection.
- Line 27 sets up the SQL statement.
- Line 29 executes the statement.
- Lines 30 through 31 close the database.
- Lines 37 through 69 are the dumpTblEntrancelog routine that follows the same pattern of opening and closing the database, except that the SQL string is a query string that dumps the Entrance Log values from the database.
- Lines 73 through 104 are the dumpTblEmployee routine that follows the same pattern of opening and closing the database as described for saveData routine. The SQL string is a query string that dumps the Employee values from the database.

LISTING 6.1 /book/chapter6/test.c program demonstrates the database interaction.

```
1)    #include "/usr/include/pgsql/libpq-fe.h"
2)    #include <stdio.h>
3)    #include <stdlib.h>
4)    #include <fcntl.h>
```

```
5)    #include <unistd.h>
6)    #include <string.h>
7)    #include <ctype.h>
8)    #include <errno.h>
9)    #include <sys/types.h>
10)   #include <sys/stat.h>
11)   #include <sys/ioctl.h>
12)   PGconn *conn;
13)   PGresult *res;
14)   /**************************************/
15)   /* Save the string into the database */
16)   /**************************************/
17)   int saveData(char * id_string)
18)   {
19)      char query_string[80];
20)      conn = PQconnectdb("dbname=timelog");
21)      if (PQstatus(conn)==CONNECTION_BAD)
22)      {
23)         fprintf(stderr,"ERROR ****** Connection to Database failed\n");
24)         exit(1);
25)      }
26)      id_string[4]='\0';
27)      sprintf(query_string,"Insert into entrancelog (id, inout) values
         ('%s',       'IN');",id_string);
28)      printf("%s\n",query_string);
29)      res = PQexec(conn,query_string);
30)      PQclear(res);
31)      PQfinish(conn);
32)      return 0;
33)   }
34)   /**************************************/
35)   /* Dump the contents of database     */
36)   /**************************************/
37)   int dumpTblEntrancelog(char * id_string)
38)   {
39)      int i;
40)      char query_string[80];
41)      conn = PQconnectdb("dbname=timelog");
42)      if (PQstatus(conn)==CONNECTION_BAD)
43)      {
44)         fprintf(stderr,"ERROR ***Connection to Database failed\n");
45)         fprintf(stderr,"Root Permission not allowed\n");
46)         exit(1);
47)      }
```

```
48)     sprintf(query_string,"SELECT * FROM entrancelog Where
        id='%s';",id_string);
49)     printf("%s\n",query_string);
50)     res = PQexec(conn,query_string);
51)     if (PQresultStatus(res) == PGRES_TUPLES_OK)
52)     {
53)        printf("ID—-TimeStamp—InOut\n");
54)        for (i=0; i < PQntuples(res); i++)
55)        {
56)           printf("%s  %s %s\n",
57)           PQgetvalue(res,i,0),
58)           PQgetvalue(res,i,1),
59)           PQgetvalue(res,i,2));
60)        }
61)     }
62)     else
63)     {
64)        fprintf(stderr,"Query failed\n");
65)     }
66)     PQclear(res);
67)     PQfinish(conn);
68)     return 0;
69)  }
70)  /*************************************/
71)  /* Dump the contents of database     */
72)  /*************************************/
73)  int dumpTblEmployee()
74)  {
75)     int i;
76)     char query_string[80];
77)     conn = PQconnectdb("dbname=timelog");
78)     if (PQstatus(conn)==CONNECTION_BAD)
79)     {
80)        fprintf(stderr,"ERROR ***Connection to Database failed\n");
81)        fprintf(stderr,"Root Permission not allowed\n");
82)        exit(1);
83)     }
84)     sprintf(query_string,"SELECT * FROM employee;");
85)     res = PQexec(conn,query_string);
86)     if (PQresultStatus(res) == PGRES_TUPLES_OK)
87)     {
88)        printf("ID —-Password —- Name\n");
89)        for (i=0; i < PQntuples(res); i++)
90)        {
```

```
91)         printf("%s  %s %s\n",
92)         PQgetvalue(res,i,0),
93)         PQgetvalue(res,i,1),
94)         PQgetvalue(res,i,2));
95)       }
96)     }
97)     else
98)     {
99)       fprintf(stderr,"Query failed\n");
100)    }
101)    PQclear(res);
102)    PQfinish(conn);
103)    return 0;
104) }
105) /******************************/
106) int main()
107) {
108)    char id[80];
109)    int i;
110)    while (1)
111)    {
112)      printf ("Please make selection \n");
113)      printf ("1 - enter timestamp \n");
114)      printf ("2 - dump the employee table\n");
115)      printf ("3 - dump the entrancelog table\n");
116)      printf (" Any other digit to exit \n");
117)      scanf ("%i",&i);
118)      switch(i)
119)      {
120)      case 1: printf ("Please enter 4 digit ID ");
121)              scanf ("%s", id);
122)              saveData(id);
123)              break;
124)      case 2: dumpTblEmployee(id);
125)              break;
126)      case 3: printf ("Please enter 4 digit ID ");
127)              scanf ("%s", id);
128)              dumpTblEntrancelog(id);
129)              break;
130)      default:
131)        exit(0);
132)      }
133)    }
134) }
```

Makefile

Use the following Makefile in Listing 6.2 /book/chapter6/Makefile to compile and link the test.c program in Listing 6.1.

LISTING 6.2 Makefile for test.c.

```
all: t
CC = gcc
CFLAGS = -g -O2 -Wall -I/usr/include -I/usr/include/pgsql
t: test.o
	$(CC) -o t test.o -lpq -lcrypt
test.o: test.c
	$(CC) -c test.c $(CFLAGS)
```

SUMMARY

In this chapter, we discussed the relational database concept and defined the basics of the Structured Query Language (SQL) for interacting with database management systems. We mentioned the different SQL statements that affect the process of table creation (CREATE), insert values into the tables (INSERT), and remove the table from the database (DROP). We also discussed SQL queries as a method of extracting information from the database. We discussed PostgreSQL from an application point of view, and implemented a simple C language application to demonstrate the different function calls to interface with the database.

7 Shell Script Programming

In this chapter

- Shell Programming
- Variables
- Conditional Statements
- System Command Expansion
- Built-in Commands
- Pipelines

INTRODUCTION

At your command—that is what a shell is, waiting for your command. You type in your command and press Return; the shell will analyze and interpret the line you typed and execute the command. Working with Linux requires understanding shell programming, as scripts for starting and stopping a process are written as shell program statements, commonly known as *shell commands.*

The general format for a command is:

```
command argument
```

The arguments are essentially parameters to the function requested in the command, but not all commands need arguments. The operating system opens a shell for the user to interact with and provides a set of functionality. The following are basic functions provided by the shell:

- Program execution
- System environment control
- Interpret as a programming language
- Filename substitution
- I/O redirection
- Pipeline input and output stream

The keyboard is not the only way to submit commands; if your command list is long, you can enter them in a file and let the shell read it and execute it for you—which is the main topic of discussion in this chapter. A file that can be interpreted by a shell is called a *shell script*. There is no difference between the command lines in a shell script and commands entered via the keyboard. The only thing you need to add in the shell script is a line at the top that tells what type of shell should interpret the rest of the lines. There are several different shells available on Linux; for example, bash, csh, ksh, and so forth. The differences among them are subtle, so the script you write based on one shell might not be compatible with the others. The syntax presented in this chapter is mainly written for bash (Bourne Again Shell). When bash is invoked as an interactive login shell, or as a noninteractive shell with the —login option, it first reads and executes commands from the file /etc/profile, if that file exists. After reading that file, it looks for ~/.bash_profile, ~/.bash_login, and ~/.profile, in that order, and reads and executes commands from the first one that exists and is readable. The —noprofile option can be used when the shell is started to inhibit this behavior.

SHELL PROGRAMMING

Shell provides a complete programming environment; there are variables, statements, expressions, substitutions, loop structures, conditional statements, and subroutines. You can also access built-in functions of shell, and pipe the output of one command into the input of another. Linux depends on shell for most of its user interface as well as system setup. The boot process of Linux executes shell scripts to start its services as we saw in the section *Linux Boot Process* in Chapter 1. This chapter provides a tutorial for shell script programming.

Setting the File Permission's Executable Bit

A file that you want a shell to execute must have the executable bit set. You can view a file's permission with the following command:

```
$ ls –l myfile
-rwxr-x—x  1 root  root   2001 jul 18 myfile
```

The file permissions are displayed as three separate fields of three bits each, preceded by – if it is a file or d if it is a directory. The fields are r (read), w (write), and x (executable). The first three bits are *owner* permissions, the middle three are *group* permissions, and the last three are *other* permissions. In the preceding example, the file "myfile" has *owner* permission rwx, the *group* permissions r-x, and *other* permissions are —x. The owner can read, write, and execute; the group can only read and execute; and others can only execute. A "–" means permission is denied.

You can make a file executable with the "chmod" command as follows:

```
$ chmod a+x myfile
```

The following command will enable all the permission bits, thus making a file read, write, and executable by everybody (not a good practice):

```
$ chmod 777 myfile
```

Making the File Interpretable

The first line of the shell script informs which shell should be invoked to interpret the rest of the lines. The following line of text in a file indicates that the file should be interpreted by the shell program /bin/bash. The special format for the line is #! followed by the executable path.

```
#!/bin/bash
```

Although the "#" character in the first column is reserved for comments, the "#" following by "!" invokes the executable filename.

Comments in a Shell Script

The lines preceded by "#" are treated as comments, except for #!.

```
#This is a comment
```

A Simple Shell Script

Edit a file named "myfile", and enter the following text in it:

```
#!/bin/bash
    print 'Hello World'
```

Make the File Executable

```
$ chmod a+x myfile
```

Execute the file and see the response "Hello World."

```
$ ./myfile
Hello World
```

VARIABLES

You do not need to declare variables. You create and assign values to them, as you need them. Simply enter a variable name followed by "=", followed by the parameter as ASCII string. A numeric variable is treated like a string variable. Type checking will be done at the time of operations performed on the variables.

```
$ testing=Hello
```

When you use the variable, you must precede it with the $ sign.

```
$ echo $testing
Hello
```

You must quote the variables if there are spaces in the parameters.

```
$ testing="Hello World"
$ echo $testing
Hello World
$ testing="1 + 3"
$echo $testing
1 + 3
```

Quotations

The behavior of the variable inside the quotes depends on the type of quotation marks you use.

Double Quotes

If you enclose a $ variable in double quotes, the variable is replaced with its value.

```
$ echo "$testing"
1 + 3
```

No substitution takes place if you enclose them in single quotes.

```
$ echo '$testing'
$testing
```

A character preceded with a "\" will be treated literally.

```
$ echo \$testing
$testing
```

Using Shell Variables

You can assign a value to a variable programmatically. The read command gets input from the keyboard.

```
$ read testing
$ TTTT
$ echo $testing
TTTT
```

Environmental Variables

When a shell is invoked, it inherits some of its variables from the /etc/.cshrcconf file. The following is a list of variables that are created when the shell is invoked the first time:

$HOME	The path to the directory.
$PATH	The colon-separated list of directories to search for commands.
$PS1	A command prompt.
$PS2	A secondary prompt for additional input.
$IFS	Input field separator. A list of characters that are used to separate words when the shell is reading input.
$0	The name of the shell script.
$#	The number of parameters passed.
$$	The process ID of the shell script.

Parameter Variables

If your script is invoked with parameters, some additional variables are created. The parameters are essentially expanded before they are referenced in a command line.

No values can be assigned to these parameters. The following is a list of special variables and the explanation of expansion being performed by the shell:

$1 $2 $3 The parameters given to the script.

$IFS Input field separator.

@ Expands the input positional parameters. "$@" is equivalent to "$1" "$2"; if there are no positional parameters, "$@" and $@ expand to nothing (i.e., they are removed).

* Expands similar to @, except it expands to a single word with the value of each parameter separated by the first character of the IFS special variable. That is, "$*" is equivalent to "$1c$2c...", where c is the first character of the value of the IFS variable. If IFS is unset, the parameters are separated by spaces. If IFS is null, the parameters are joined without intervening separators.

Expands to the number of positional parameters in decimal.

? Expands to the status of the most recently executed foreground pipeline.

- Expands to the current option flags as specified upon invocation, by the set built-in command, or those set by the shell itself (such as the –I option).

$ Expands to the process ID of the shell. In a () subshell, it expands to the process ID of the current shell, not the subshell.

! Expands to the process ID of the most recently executed background (asynchronous) command.

0 Expands to the name of the shell or shell script. This is set at shell initialization. If bash is invoked with a file of commands, $0 is set to the name of that file. If bash is started with the -c option, then $0 is set to the first argument after the string to be executed, if one is present. Otherwise, it is set to the filename used to invoke bash, as given by argument zero.

_ At shell startup, set to the absolute filename of the shell or shell script being executed as passed in the argument list. Subsequently, expands to the last argument to the previous command, after expansion. Also set to the full filename of each command executed and placed in the environment exported to that command. When checking mail, this parameter holds the name of the mail file currently being checked.

The following example is a command and response sequence to illustrate the effects of the special parameter expansion:

```
$ set buck stop here
$ echo "$@"        #replace @ with its value
buckstophere
$ echo "$*"                #replace the variable $* with its value
buck stop here
$ IFS=''                   #empty separator
$ echo "$*"                #replace the variable $* with its value
buckstophere               #notice the separator is empty
$ unset IFS                #bring back the original separator
$ echo "$*"
buck stop here             #back with the space as separator
```

Arrays

Bash allows single-dimensional arrays; to declare a variable as an array, use the syntax:

```
name[subscript]=value.
```

Arrays are indexed using integers and are zero-based. Compound values can be assigned to arrays using the form:

```
name=(value1 ... valuen)
```

Any element of an array can be referenced using ${name[subscript]}. The braces are required to avoid conflicts with the pathname expansion. Referencing an array variable without a subscript is equivalent to referencing element zero. Please see the section *Expansion Command* for a complete explanation of expansion.

The *unset* is used to destroy arrays. Unset name[subscript] destroys the array element at index subscript unset name, where name is an array, or unset name[subscript], where subscript is * or @, removes the entire array.

CONDITIONAL STATEMENTS

All programming languages provide some form of test and condition statements. In the shell, the *if* statement performs this function, and the word *test* followed by an expression provides the condition statement. Enter the following commands at the shell prompt and notice the change of prompt until the if block is complete:

```
$ testing=1
$ if test $testing
>then
```

```
>echo "true"
>fi
true
```

A better way to express the test command is by embedding the condition between the brackets [ ]. The following statements are the same:

```
if test $testing
if [ $testing ]
You must provide space after the opening #bracket '[' and before the
closing bracket ']',
$ if [ $testing ]
>then
>echo "true"
>fi
true
```

If you want to add more statements followed by the closing bracket, you must end the condition statement with the ";".

```
$ if [ $testing ] ; then
>echo "true"
>fi
true
```

The condition types supported by the shell command fall into the following three categories discussed next.

String Comparison

String	True if the string is not an empty string.
String1 = string2	True if the strings are the same.
String1 != string2	True if the strings are not the same.
-n string	True if the string is not null.
-z string	True if the string is null (an empty string).

For example:

```
$ testing1="hello world"
$ testing2="hello"
$ if [ $testing1 = $testing2 ]; then echo "true"
>else echo "false"
false
```

The shell scripts can test the *exit* code of the other command and shell scripts that can be invoked from the shell command. That is why it is important to include an exit statement within a shell script.

Arithmetic Comparison

Expression1 –eq Expression2	True if the expressions are equal.
Expression1 –ne Expression2	True if the expressions are not equal.
Expression1 –gt Expression2	True if expressions1 is greater than expression2.
:Expression1 –ge Expression2	True if expressions1 is greater than or equal to expression2.
Expression1 –lt Expression2	True if expressions1 is less than expression2.
Expression1 –le Expression2	True if expressions1 is less than or equal to expression2.
!Expression	True if expression is false (! is the negate operation).

File Conditions

The shell provides special test conditions for files and directories. These are handy when setting up the startup script, as some programs depend on the presence of a configuration file for proper execution.

-d file	True if the file is a directory.
-e file	True if the file exists.
-f file	True if the file is a regular file.
-g file	True if set-group-id is set on the file.
-r file	True if the file is readable.
-s file	True if the file is a nonzero size.
-u file	True if set-user-id is set on the file .
-w file	True if the file is writeable.
-x file	True if the file is executable.

The following lines are from the shell script network from /etc/rc.d/init.d/network. The command checks for the presence of the file pcmcia and executes it if it is present.

```
if [ -f /etc/sysconfig/pcmcia ]; then
    . /etc/sysconfig/pcmcia
fi
```

SYSTEM COMMAND EXPANSION

Expansion is performed on the command line after it has been split into words. The shell performs the following types of expansion on the command line:

- Brace expansion
- Tilde expansion
- Parameter and variable expansion
- Arithmetic expansion
- Process substitution
- Command substitution
- Word splitting
- Pathname expansion

Brace Expansion

Multiple strings are generated with expansion; for example, a{d,c,b}e expands into "ade ace abe". Brace expansion is performed before any other expansions, and any characters special to other expansions are preserved in the result. The following command will create two directories, old and new:

```
mkdir /root/{old, new}
```

Tilde Expansion

Generally, the "~" is replaced with the home directory associated with the specified login name. The following command will change the directory to root if you are a root user:

```
cd ~
```

$ Parameter Expansion

The "$" character introduces parameter expansion, command substitution, or arithmetic expansion. The parameter name or symbol to be expanded may be enclosed in braces. The braces are required when the parameter is a positional parameter with more than one digit, or when the parameter is followed by a character that is not to be interpreted as part of its name. The following command will substitute the value of the parameter:

```
${parameter}
```

The different form of parameter expansion is explained here:

${param:-default}	If param is null, set it to the value of default.
${#param}	Gives the length of the param.
${param%word}	From the end, remove the smallest part of the param that matches word, and return the rest.
${param%%word}	From the end, remove the longest part of the param that matches word, and return the rest.
${param#word}	From the beginning, remove the smallest part of the param that matches word, and return the rest.
${param##word}	From the beginning, remove the longest part of the param that matches word, and return the rest.

Command Substitution

Command substitution allows the output of a command to replace the command name. There are two forms:

```
$(command)
```

or

```
'command'
```

Bash performs the expansion by executing command and replacing the command substitution with the standard output of the command, with any trailing newlines deleted. Embedded newlines are not deleted, but they may be removed during word splitting. The command substitution $(cat file) can be replaced by the equivalent but faster $(<file).

[]

The command $(ls –al t[345]) in effect expands to"

```
$ ls -al t3
$ ls -al t4
$ ls -al t5
*
```

The '*" is treated as a wild card character that matches to all combinations. The command $(ls –al t[*]) in effect expands to all combinations of characters that begin with "t".

```
$ ls -al t1
$ ls -al t11
$ ls -al t111
```

Arithmetic Expansion

Arithmetic expansion allows the evaluation of an arithmetic expression and the substitution of the result. The format for arithmetic expansion is:

```
$((expression))
```

The expression is treated as if it were within double quotes, but a double quote inside the parentheses is not treated specially. All tokens in the expression undergo parameter expansion, string expansion, command substitution, and quote removal. Arithmetic substitutions can be nested.

```
$   x=0
$   while [ "$x" -ne 5 ]; do
>   echo $x
>   x=$((x+1))
>   done
0
1
2
3
4
```

Process Substitution

Process substitution is supported on systems that support named pipes (FIFOs) or the /dev/fd method of naming open files. It takes the form of <(list) or >(list). The process list is run with its input or output connected to a FIFO or some file in /dev/fd. The name of this file is passed as an argument to the current command as the result of the expansion. If the >(list) form is used, writing to the file will provide input for list. If the <(list) form is used, the file passed as an argument should be read to obtain the output of list. When available, process substitution is performed simultaneously with parameter and variable expansion, command substitution, and arithmetic expansion.

Word Splitting

The shell treats each character of IFS as a delimiter, and splits the results of the other expansions into words on these characters. If IFS is unset, or its value is exactly <space><tab><newline>, the default, then any sequence of IFS characters serves to delimit words. Note that if no expansion occurs, no splitting is performed.

Pathname Expansion

After word splitting, unless the -f option has been set, bash scans each word for the characters *, ?, and [. If one of these characters appears, the word is regarded as a pattern, and replaced with an alphabetically sorted list of filenames matching the pattern. The special pattern characters have the following meanings:

- * Matches any string, including the null string.
- ? Matches any single character.
- [...] Matches any one of the enclosed characters.

Control Structure

if

You have already seen one example of the control statement in the form of an "if" statement. Shell programming has the following format for "if" conditionals:

```
if
condition
then
statement
else
statement
fi
```

The following commands are from the script /etc/rc.d/init.d/network. The script tests to see if linuxconf file is present and executable. If it is, the file is invoked with the parameter –himt netdev; otherwise, the script executes ifconfig and pipes the output to the grep utility that prints all the lines that begin with the characters a–z.

```
if [ -x /sbin/linuxconf ] ; then
    eval `/sbin/linuxconf –hint netdev`
    echo $"Devices that are down:"
    echo $DEV_UP
    echo $"Devices with modified configuration:"
    echo $DEV_RECONF
else
    echo $"Currently active devices:"
    echo `/sbin/ifconfig | grep ^[a-z] | awk '{print $1}'`
fi
```

for

The "for" construct helps repeat the statement until all values are exhausted. For example:

```
$ for try in how are you
>do
>echo $try
>done
how
are
you
```

The parameter list for the "for" command can be the output of a system command, such as $(command). For example, the output of the list command "ls" can be used as a parameter list.

```
$ for try in $ ( ls )
>do
>echo $try
>done
```

You can use Shell expansion command as a parameter list

```
$ for try in $ ( ls –al t[345])
>do
>echo $try
>done
```

The syntax is simply

```
for variable in values
do
statement
done
```

while

The “while” control command repeats the statements between do and done until the condition becomes false.

```
    $       $myvar=1
$   While [ "$myvar" <= "8" ]; do
    >       Echo "$myvar"
>   $myvar=$((myvar+1))
>   done
```

The syntax is

```
while condition do
statements
done
```

until

The “until” control command repeats the statements between do and done until the condition becomes true (opposite of while).

```
$   $myvar=1
$   until [ "$myvar" > "8" ]; do
    >       Echo "$myvar"
>   $myvar=$((myvar+1))
>   done
```

The syntax is

```
until condition do
statements
done
```

Case Statement

The "case" control command takes the content of the variable and compares it to each pattern string in turn. If a string matches, the statement after the ")" is executed. The following is the format for case block:

```
case variable in
    Pattern ) statements;;
    Pattern ) statements;;
    Pattern ) statements;;
esac
```

The shell compares the input parameter with "start" and executes the block if a match found.

```
if case "$1" in
 start)
       action $"Setting network parameters: " sysctl -e -p /etc/sysctl.conf
       action $"Bringing up interface lo: " ./ifup ifcfg-lo case "$IPX" in
yes|true)
       /sbin/ipx_configure —auto_primary=$IPXAUTOPRIMARY \
                                      —auto_interface=$IPXAUTOFRAME
 esac
esac
```

Lists

If you want to connect commands in a series, you have two options, as discussed in the following sections.

The AND Lists

```
If [ condition1 ] && statement1 [ condition2 ] && [ condition3 ] ;
statement2;
```

The statement1 is executed if condition1 is true. The statement2 is executed if condition1, condition2, and condition3 are true.

The OR Lists

```
If [ condition1 ] || statement1 [ condition2 ] || [ condition3 ] ;
statement2;
```

In this case, the rule for executing the next statement is that the preceding condition must fail.

Statement Block

All statements inside the braces {} are treated as one entity. This is how multiple statements are executed where only one is allowed. For example:

```
If [ condition1 ] || {
statement1
statement2
statement3
}
[ condition2 ] || [ condition4 ] ; statement5;
```

If condition1 is false, then all the statements inside the braces (statement1, statement2, and statement3) are executed.

Functions

A function is a group of statements and commands enclosed in braces that are assigned a unique name followed by (). Functions are necessary, as it helps to confine a section of code that is repeated many times in the main program block, which helps to make the script readable. The following function takes two integer arguments and prints the result of comparison:

```
1)  #!/bin/sh
2)  func1 () {
3)  if [ $1 -le $2 ]
4)  then
5)   echo "$1 is less then or equal to $2 "
6)  else
7)   echo "$1 is greater then $2 "
8)  fi
9)  }
10) func1 2 3
11) func1 6 5
12) 2 is less then or equal to 3
13) 6 is greater the 5
```

- Line 1 is the usual script invocation.
- Line 2 is the function declaration with the name followed by ().
- Line 3 is the beginning of the if block with test condition of comparing the two input variables.

- Lines 4 and 5 are the true block that is executed when the result is true.
- Lines 6 and 7 are the false block that is executed when the result is false.
- Line 8 is the end of the if block, and line 9 is the closing of the function block.
- Line 10 is the function call, passing arguments 2 and 3.
- Line 11 is the function call, passing arguments 6 and 5.
- Lines 12 and 13 are the shell response.

Once a function is defined, you can call the function like a statement or inside a condition construct. The return value of the function is the last statement executed inside the function, or you can use a return statement with a numeric value. Returning a string value is possible by assigning a variable that has the same scope as the section of the code

Passing Arguments to Functions and Programs

Function names are followed by an argument list when a function is being called. However, the argument names are not visible to the function being called if the function scope is outside the variable. Instead, the shell copies the arguments to special variables before a function is being called. These variables are $1, $2, $3, and so forth, where $1 is the first variable, $2 is the second variable, and so on. The following example demonstrates the use of $n variable. The function echoes the argument that is being passed.

```
    $ echoarg()
    $ {
    $ echo $1
$ }
$ echoarg "Try this"
Try this
```

BUILT-IN COMMANDS

There are two types of commands in a shell. The normal commands that we have discussed before, like "ls", that can be executed on the shell command line, and the built-in commands that are implemented internally and cannot be invoked as external programs.

break

Break is used in early termination of the loop construct of for, while, or until, before the controlling condition has been met.

The : Command

The colon ":" is a null command. It is also an alias for true. The while : implements an infinite loop.

continue

Continue makes the for, while, and until loop continue at the next iteration.

The . Command

The . command executes the command in the current shell.

echo

The shell equivalent of printf command. Anything followed by echo is printed as is, except the "$" sign, "\", and quotes. The word after the "$" sign is treated as a variable unless you enclose the entire string in single quotes. The character followed by the "\" is treated literally; for example, \$try is printed as $try.

eval

The eval command allows you to evaluate arguments.

exec

Exec is normally used for replacing the current shell with a different program. The other use of exec is to modify a file descriptor.

exit n

The exit command causes the script to exit with exit code n.

export

The export command makes the variable named as a parameter available in subshells.

expr

The expr command evaluates its arguments as an expression. The following is a list of expr commands:

expr1 | expr2	expr1 if expr1 is nonzero; else, expr2
expr1 & expr2	Zero if either expression is zero; otherwise, expr1
expr1 = expr2	Equal

expr1 > expr2	Greater than
expr1 >= expr2	Greater than or equal
expr1 < expr2	Less than
expr1 <= expr2	Less than or equal
expr1 != expr2	Not equal
expr1 + expr2	Add
expr1 - expr2	Subtract
expr1 * expr2	Multiply
expr1 / expr2	Divide
expr1 % expr2	Divide modulus

printf

Same as echo.

return

The return command causes the function to return. It can take a single numeric parameter. If no parameter is provided, the result of the last statement executed will be returned.

set

Sets the parameter value of the shell.

shift

Shift command moves all parameter variables down by one; $2 becomes $1, $3 becomes $2, and so on. The $1 value is discarded.

trap

The trap command is used for specifying the actions to take on receipt of a signal. The most common signals are:

HUP(1)	Hang up, usually sent when the terminal goes offline
INT(2)	Interrupt, usually Ctrl C
QUIT(3)	Quit, usually sent by Ctrl \
ABRT(4)	Abort, usually sent on some serious error
TERM(15)	Terminate, usually sent by system shutdown

unset

The unset command removes the variable from the environment or function.

PIPELINES

A pipeline is a sequence of one or more commands separated by the "|"character. The pipe command essentially connects the standard output of one command to the standard input of another. The shell opens a subshell for each command in the pipeline. The following example displays all files that have the word *X11*, starting from the current directory:

```
find . | grep X11
```

Redirections and Duplicating File Descriptors

There are three default file descriptors associated with the standard input, output, and error devices when a shell is opened for command interpretation. Descriptor 0 is the keyboard input, 1 is the screen output, and 2 is the standard error output, which is also directed to screen. The input, output, and error can be redirected using a special notation interpreted by the shell. Redirection can also be used to open and close files for the current shell execution environment. Redirections are processed in the order in which they appear, from left to right.

The following redirection operator is used to duplicate the input file descriptors:

```
[n]<&word
```

Similarly, the following operator duplicates the output file descriptor:

```
[n]>&word
```

If n is not specified, the standard output (file descriptor 1) is used.

The following redirection operator opens file descriptors for reading and writing:

```
[n]<>word
```

< Redirecting input refers to the standard input (file descriptor 0). The input is taken from a file.

> Redirecting output refers to the standard output (file descriptor 1). Output is redirected to a file.

>> The output is appended at the end of the file.
For example

```
ls > dirlist 2>&1
```

redirects both standard output and standard error to the file dirlist, while the command

```
ls 2>&1 > dirlist
```

redirects only the standard output to file dirlist, because the standard error was duplicated as standard output before the standard output was redirected to dirlist. There are two formats for redirecting standard output and standard error:

```
&>word
```

and

```
>&word
```

Of the two, the first is preferred. This is equivalent to

```
>word 2>&1
```

A Here Document

A here document starts with a leader << followed by a special sequence of characters that will be repeated at the end of the document. << is a label redirector.

```
cat <<++TTT++
Hello
How are you
++TTT++
```

The cat command gets redirected input from the keyboard instead of a file. The special sequence starts with ++TTT++ and ends with ++TTT+++.

SUMMARY

In this chapter, we discussed Linux shell command processing from a programming point of view. The executables (*shell scripts*, as they are called) are extensively used in the startup of different daemons in the operating system. Shell programming's unique features are command substitution and file redirections.

8 Fixed-Point Arithmetic and Transcendental Functions

In this chapter

- Transcendental Functions
- Reduced Coefficient Polynomials
- E^x (Exponent of X)
- Transcendental Functions Library

INTRODUCTION

Not all microprocessors are created equally; some have math coprocessors and some do not. In fact, the Intel 80486 was the first Intel microprocessor that had a built-in math coprocessor; otherwise, up until 80386, you had to pay extra for any decent processing power. C compilers were provided with the math emulation routines, but the performance was anything but acceptable. It was not because the routines were written poorly; there was a legacy and a standard for the floating-point arithmetic. They all had to confirm to IEEE standards for double precision arithmetic of 64-bit words; especially, the solution of transcendental functions. The sin, cos, tan, exp, logarithm, and so forth were all written with a high degree of accuracy, but a penalty was paid in terms of performance. Graphics routines were the hardest hit. The transformation and rotation of objects requiring matrix multiplication and additions were noticeably slow, mainly because the underlying arithmetic operations were slow. Today, we take math coprocessors for granted, but there are still times when your embedded system design cannot afford a coprocessor and you have to come up with a fast and efficient math library. This chapter is devoted to math algorithms and solution of transcendental functions. The fixed-point arithmetic is presented as an alternative to floating-point arithmetic, and the pros and cons are discussed with examples. One more point to highlight is the magnitude of the work involved in the solution of transcendental functions, and you can make an informed decision regarding your need for a math coprocessor. We begin our discussion with the solution of transcendental functions.

TRANSCENDENTAL FUNCTIONS

By definition, the trigonometric and logarithm functions, including *sin*, *cos*, *tan*, *e*, *and ln*, and so forth, cannot be expressed in algebraic form. Just like the square root of 2 and the value of pi, these are not rational numbers, which means that there are no two integers whose ratio can be used to define them. As we know, all arithmetic operations must be brought down to the four basic operations of addition, subtraction, multiplication, and division. If a function cannot be expressed as an operation on rational numbers, it cannot be solved. Thus, the only method of computing the transcendental functions is by the expansion of the function in its polynomial form. The Taylor series expansion is the fundamental solution to all transcendental functions.

The following is an example of the Taylor series expansion for the exponent function:

$$e^x = 1 + {x^1}/{1!} + {x^2}/{2!} + {x^3}/{3!} + {x^4}/{4!} + {x^5}/{5!} + \ldots\ldots + {x^n}/{n!}$$

One must expand the series to its infinite terms if an exact solution of the function is required. Obviously, nobody has the time and, probably, nobody needs the precision either. So, how do you decide how many terms of the series should be used? That depends on your application and the degree of precision you need. If you want to find the distance between the stars, go for a higher precision, but if you only need to display a picture on a 800x600-pixel computer monitor, all that precision computation is a waste—if you are computing the solutions by hand or using a software algorithm. The precision computation done with software has a hidden cost; with every gain of precision improvement, the processing time increases.

Unfortunately, you might not notice the degradation of software performance if you are using the standard C library routines. The routines are basically "one size fits all." We cannot selectively increase or decrease the degree of the polynomials that are being used to compute the underlying functions. The standard calls for simply a double-precision arithmetic of 64-bits, comprising 11 bits of exponent, 53 bits of mantissa or fraction, and 1 sign bit (the total is 65 bits, we gain an extra bit for using the normalized mantissa as explained later in this section). Figure 8.1 describes the format for a double-precision floating-point number. The float variable, on the other hand, is a 32-bit-long word with 1 sign bit, 8 exponent bits and 24 bits of mantissa or fraction. Figure 8.2 describes the format for a single-precision number. Again, we gain an extra bit of storage by removing the normalized bit of mantissa and using it to store an extra precision bit.

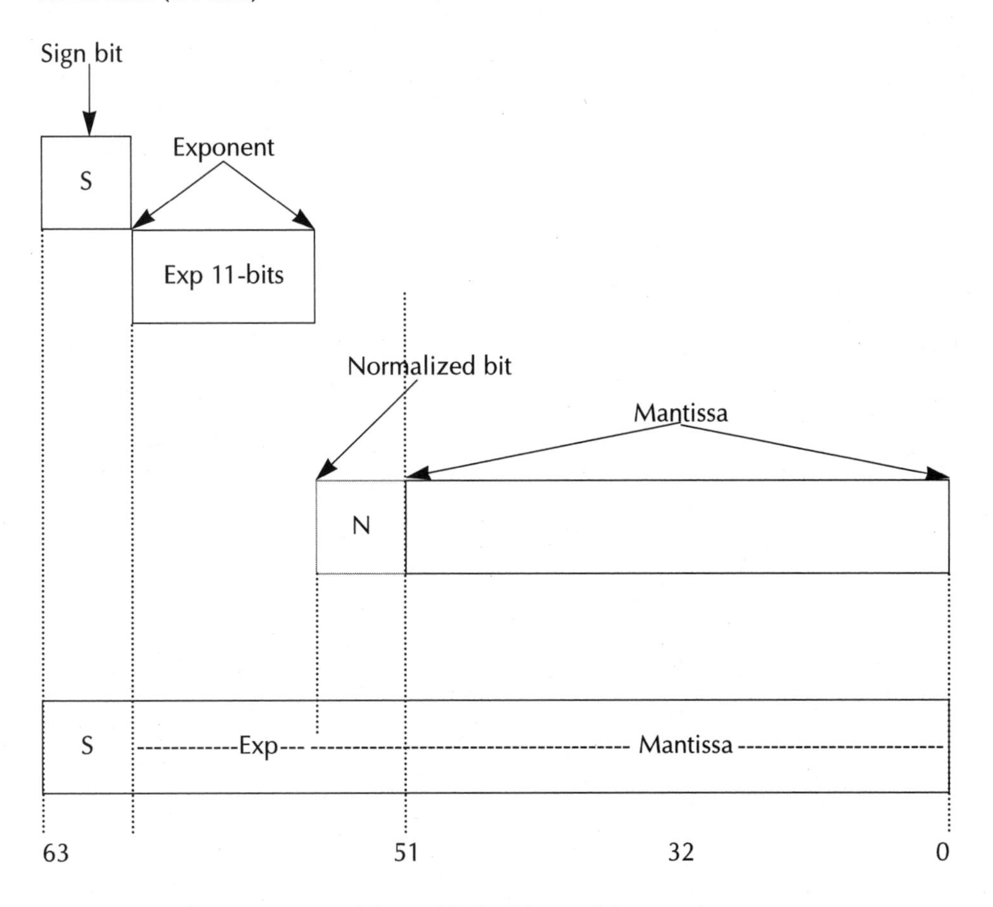

FIGURE 8.1 The components of a 64-bit double precision word.

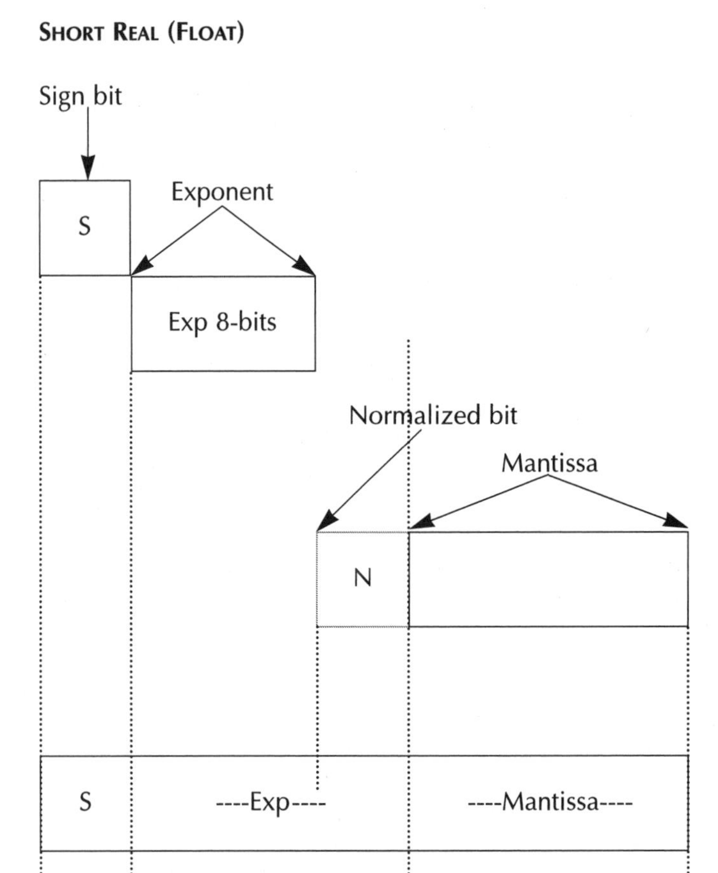

FIGURE 8.2 The components of a 32-bit single-precision word.

You might think that specifying a variable as "float" instead of "double" might give you performance improvement, as there are only 32 bits in a float variable—think again! It is true that multiplying two 32-bit numbers is faster than multiplying two 64-bit numbers, but all numbers are converted to 64-bit double precision, before the polynomial operation being performed. This is double jeopardy—now there is the extra penalty of converting the numbers from single precision to double precision, and vice versa.

If your embedded system cannot afford a math coprocessor and you need floating-point arithmetic, the following is an alternate solution to the standard C library functions. The precision is not that great—in some cases, it is about four decimal places—but you will find these routines several times faster than the standard functions, and you might not need all that extra precision after all.

There is a two-pronged approach that is used in the upcoming implementation. First, we have selected the lower-degree polynomials that reduce the amount of multiplication operations. Second, all the basic arithmetic operations of multiply, divide, add, and subtract are done using the 32-bit integer arithmetic.

We will begin our discussion with the general solution of transcendental functions; in other words, polynomial expansion.

REDUCED COEFFICIENT POLYNOMIALS

A polynomial is a truncated form of an infinite series, suitable for computational purposes. We must stop solving the terms after achieving a certain degree of precision, as we cannot go on computing forever, as required for an infinite series. If we were to compute the value of the exp function with the help of the Taylor series expansion as explained previously, we need to evaluate the terms for at least 100 degrees before we get a result of three decimal places of accuracy—not a very practical suggestion. What we need is to combine the coefficients beforehand in order to reduce the degree of the terms in our final analysis. B. Carlson and M. Goldstein in 1943 presented polynomial solutions for transcendental functions that are suitable for digital computer applications.

There are generally two sets of solutions for each basic function of sin, cos, tan, arctan, exponent, and logarithm. The high-order polynomial is for the double-precision result, and the low-order polynomial is for the single-precision result. Other functions such as cosec, cot, and sec can be evaluated with the basic trigonometric identities. The following are the two sets of reduced coefficient polynomials, obtained from the respective infinite series, that are guaranteed to be convergent for the input range specified.

sin(x)

The input to this function is the angle x in radians, whose sin is to be determined. The governing polynomial is valid only for the range $0 \le x \le \pi/2$, Thus, the operand must be reduced to the given range by applying the modulus rule for the sin function.

Low-degree polynomial for single-precision application:

$$\text{Input range} \quad 0 \le x \le \frac{\pi}{2}$$

$$\frac{\sin(x)}{x} = 1 + a_2x^2 + a_4x^4 + \xi(x)$$

$$\text{Magnitude of error} - |\xi(x)| \le 2 * 10^{-4}$$

Constants: $a_2 = -.16605$, $a_4 = .00761$

High-degree polynomial for double-precision application:

$$\text{Input range} \quad 0 \le x \le \frac{\pi}{2}$$

$$\frac{\sin(x)}{x} = 1 + a_2x^2 + a_4x^4 + a_6x^6 + a_8x^8 + a_{10}x^{10} + \xi(x)$$

$$\text{Magnitude of error} - |\xi(x)| \le 2 * 10^{-6}$$

Constants:

$a_2 = -.1666666664$, $a_8 = .0000027526$

$a_4 = .0083333315$, $a_{10} = -.0000000239$

$a_6 = -.001984090$

cos(x)

The input to this function is the angle x in radians, whose cos is to be determined. The governing polynomial is valid only for the range $0 \le x \le \pi/2$, Thus, the operand must be reduced to the given range by applying the modulus rule for the cos function.

Low-degree polynomial for single-precision application:

$$\text{Input range} \quad 0 \le x \le \frac{\pi}{2}$$

$$\cos(x) = 1 + a_2x^2 + a_4x^4 + \xi(x)$$

$$\text{Magnitude of error} - |\xi(x)| \le 2 * 10^{-4}$$

Constants: $a_2 = -.49670$, $a_4 = .03705$

High-degree polynomial for double-precision application:

$$\text{Input range} \quad 0 \le x \le \frac{\pi}{2}$$

$$\cos(x) = 1 + a_2x^2 + a_4x^4 + a_6x^6 + a_8x^8 + a_{10}x^{10} + \xi(x)$$

$$\text{Magnitude of error} - |\xi(x)| \le 2 * 10^{-6}$$

Constants:

$$a_2 = -.4999999963, a_4 = .0416666418$$

$$a_6 = -.0013888397, a_8 = .0000247609$$

$$a_{10} = -.000002605$$

tan(x)

The governing polynomial for tan(x) is defined in terms of its counterpart x*cot(x). tan(x) is obtained by the analogy tan(x) = 1/cot(x). Again, the angle x is in radians, whose tan is to be determined. The polynomial is valid only for the range $\pi/4 \le x \le \pi/2$, Thus, the operand must be reduced to the given range by applying the modulus rule for the cotangent function.

Low-degree polynomial for single-precision application:

$$\text{Input range} \quad \le \frac{\pi}{4} \le x \le \frac{\pi}{4}$$

$$\cos(x) = 1 + a_2x^2 + a_4x^4 + \xi(x)$$

$$\text{Magnitude of error} - |\xi(x)| \le 2 * 10^{-5}$$

Constants: $a_2 = -.332867$, $a_4 = .024369$

High-degree polynomial for double-precision application:

$$\text{Input range} \quad 0 \le x \le \frac{\pi}{4}$$

$$x^*\cot(x) = 1 + a_2x^2 + a_4x^4 + a_6x^6 + a_8x^8 + a_{10}x^{10} + \xi(x)$$

$$\text{Magnitude of error} - |\xi(x)| \le 2 * 10^{-6}$$

Constants:

$$a_2 = -.3333333410, a_4 = -.0222220287$$
$$a_6 = -.0021177168, a_8 = -.0002078504$$
$$a_{10} = -.0021177168$$

E^X (EXPONENT OF X)

The polynomial for e^x is defined for values less than 0.639. The exponent of larger numbers can be computed by splitting the input floating-point value into its integer and fraction portion, and applying the following rule:

$$e^x = e^{(\text{integer portion } + 0.693 + \text{fraction portion})}$$
$$= e^{(\text{integer portion})}\, 2\, e^{(\text{fraction portion})}$$

The exponent of the integer can quickly grow out of bounds. For example, the exponent of 31 is 2.9x1013; anything above can be considered infinity.

Low-degree polynomial for single-precision application:

$$\text{Input range} \quad 0 \le x \le 0.639$$
$$e^x = {}^1/_{1.0 + a_1x^1 + a_2x^2 + a_3x^3 + a_4x^4} + \xi(x)$$
$$\text{Magnitude of error} - |\xi(x)| \le 2 * 10^{-5}$$

Constants: $a_1 = -.9998684$, $a_2 = .4982926$, $a_3 = -.1595332$, $a_4 = .0293641$

High-degree polynomial for double-precision application:

$$\text{Input range} \quad 0 \le x \le 0.639$$
$$e^x = {}^1/_{1 + a_1x^1 + a_2x^2 + a_3x^3 + a_4x^4 + a_5x^5 + a_6x^6 + a_7x^7} + \xi(x)$$
$$\text{Magnitude of error} - |\xi(x)| \le 2 * 10^{-6}$$

Constants: $a_1 = -.9999999995$, $a_2 = .4999999206$, $a_3 = -.1666653019$, $a_4 = .0416573475$, $a_5 = -.0083013598$, $a_6 = .0013298820$, $a_7 = -.0001413161$

$\ln_e(x)$ Natural Logarithm

The polynomial for natural log is defined for values less than 1. The log of a number greater than 1.0 can be computed by applying the following rule.

$\ln(x^n) = n \ln(x)$

Low-degree polynomial for single-precision application:

$$\text{Input range} \quad 0 \le x \le 1.0$$

$$\ln(1 + x) = a_1x^1 + a_2x^2 + a_3x^3 + a_4x^4 + a_5x^5 + \xi(x)$$

$$\text{Magnitude of error} - |\xi(x)| \le 2 * 10^{-5}$$

Constants: $a_1 = .99949556$, $a_2 = -.49190896$, $a_3 = .28947478$, $a_4 = -.13606275$, $a_5 = .03215845$

High-degree polynomial for double-precision application:

$$\text{Input range} \quad 0 \le x \le \frac{\pi}{2}$$

$$\ln(1 + x) = a_1x^1 + a_2x^2 + a_3x^3 + a_4x^4 + a_5x^5 + a_6x^6 + a_7x^7 + a_8x^8 + \xi(x)$$

$$\text{Magnitude of error} - |\xi(x)| \le 2 * 10^{-6}$$

Constants: $a_1 = .9999964239$, $a_2 = -.4998741238$, $a_3 = .3317990258$, $a_4 = -.2407338084$, $a_5 = .1676540711$, $a_6 = -.0953293897$, $a_7 = .0360884937$, $a_8 = -.0064535442$

The preceding two sets of polynomials offer different precision results for the function they represent. The high-degree polynomial offers higher precision, but requires a greater number of computations; just imagine computing x to the power 8, 7, 6, 5, 4, 3, 2, and so on. The low-degree polynomial only requires computing maximum x to the power of 5—a magnitude of difference in computational speed. The library functions developed in this chapter were designed around the low-degree polynomials.

Now, let us discuss the inner-workings of the computations. It is not enough to reduce the degree of the polynomial, as the bulk of CPU time is actually spent performing the four basic arithmetic operations of multiply, divide, add, and subtract. If we were to use the default library functions for the basic operations, we essentially do not gain anything. The fundamental library routines will convert the float into double, perform the operation in double, convert the double back into the float, and then return. These conversions are inherent in the library and cannot be avoided. All these unnecessary conversions defeat the purpose of using the float format. The following discussion explores other possibilities of real number representation that are more efficient than the default floating-point format.

Fixed-Point Arithmetic and Solution of Transcendental Functions

Is there an alternative to the floating-point arithmetic? We need floating point to represent the huge range of numbers we deal with. Putting –127 through + 127 exponent combined with the 23 bits of precision digits can only be done in floating-point format. The other alternative is the fixed-point format. A 32-bit number can be thought of as carrying 8 bits of integer and 24 bits of fraction, and a 64-bit number can be divided into 32 bits of integer and 32 bits of fraction. However, we then limit the numbers to a very narrow range. The maximum integer value in an 8-bit integer is 255, and the maximum precision of the fraction in a 24-bit mantissa is 1 in 16 million—nearly .000001. All rational numbers are comprised of an integer portion and a fraction portion. The decimal point is placed between them to distinguish the end of the fraction from the beginning of the integer.

The positional numbering system base b defines a real number as follows:

$$a_n b^n + .. + a_2 b^2 + a_1 b^1 + a_0 b^0 + a_{-1} b^{-1} + a_{-2} b^{-2} + a_{-3} b^{-3} .. a_{-n} b^{-n}$$

For example, the decimal number 34.61 equals:

$$0x10^n + \ldots 3x10^1 + 4x10^0 + 6x10^{-1} + 1x10^{-2} + \ldots 0x10^{-n}$$

All computers have a fixed word-length limit and are optimized for fixed word calculations. Each word can be thought of as an integer, or it could be treated as a fixed-point value with a predefined integer portion and a predefined fraction portion. *The decimal point is only a perception. The computational method is the same for both, the integer as well as the fixed-point values. The only difference is the way in which the result is accumulated from the different sources where the CPU has placed the result after an arithmetic operation.*

For example, a 32-bit number when multiplied by another 32-bit number produces a 64-bit result. The 80386 CPU utilizes the EAX:EDX register combination to store the result. The EDX register contains the most significant 32 bits, while the EAX register contains the least significant 32 bits. In the context of a pure 32-bit integer operation, we ignore the result of the upper 32 bits and keep only the lower 32 bits that is in the EAX register. However, if we think of the 32-bit value as an 8-bit integer and a 24-bit fraction, the result must be obtained by combining the upper 8 bits of the EAX and the lower 24 bits of the EDX to bring the result back into 32-bit word format (Figure 8.3).

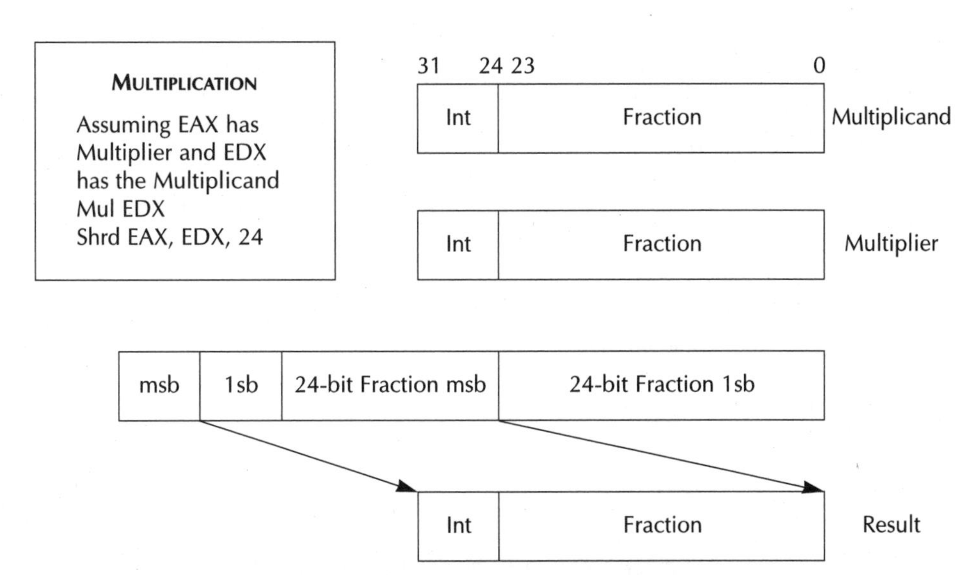

FIGURE 8.3 The multiplication of two 32-bit fixed-point numbers.

The division operation for the two 32-bit fixed-point dividend and divisor is achieved by combining the lower 8 bits of EDX and the upper 24 bits of EAX as illustrated in Figure 8.4. Notice that the 32-bit fixed-point addition and subtraction are identical to the integer addition and subtraction.

The following example demonstrates the multiplication and division operation on two 32-bit fixed-point numbers with 8 bits of integer and 24 bits of fraction.

All is well and good if we could only guarantee that the result will not exceed 255, since this is the largest number we can put in a 32-bit fixed-point format of 8-bit integer and 24-bit fraction. The fixed-point arithmetic is definitely far superior than the floating point, and allows us to perform arithmetic on real numbers with fractions—albeit, the range of numbers is very limited. Successive multiplication of two integer numbers will soon overflow the result area; moreover, successive addition might cause the carryover that has no place in the result storage. On the other hand, there is never a worry about the fraction portion being lost in the intermediate operations; at the most, we might lose some precision, which is acceptable since it happens to the floating-point format also.

If it were not for the fear of producing erroneous results in the intermediate stages, we would definitely gain considerable performance improvement by using

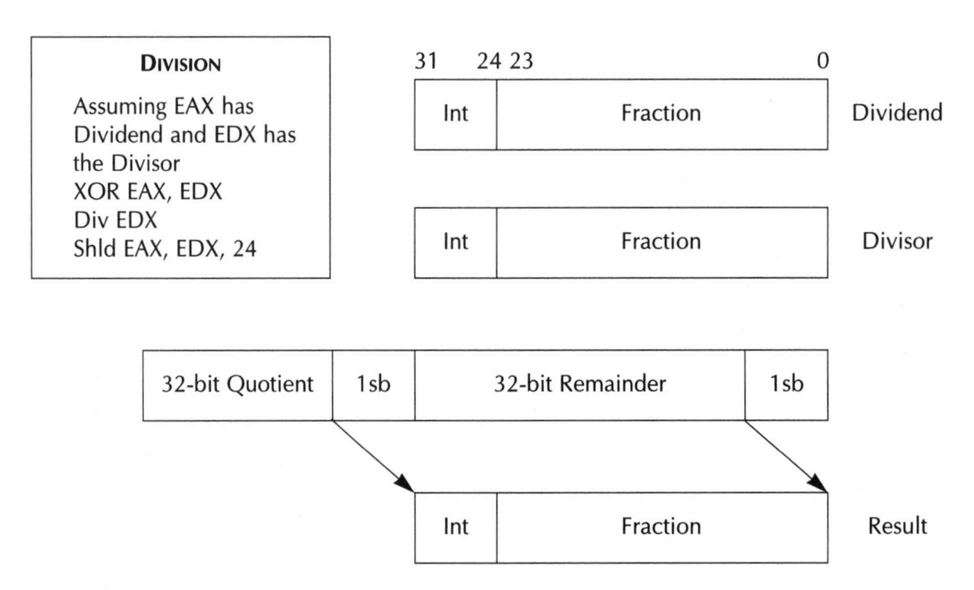

FIGURE 8.4 The division of two 32-bit fixed-point numbers.

the integer instruction set of the CPU architecture for our basic calculations on the real numbers.

Let us take another look at the arithmetic operations on the polynomial functions discussed previously and the restrictions we have on the range of the coefficients and the input values that are involved in the operation. Let us analyze all the arithmetic operations that are required to solve a given polynomial, and see if we can apply the fixed-point arithmetic instructions on the parameters. There is an input value x (always less than 1.0) that must be raised to its power several times; that is, successive multiplications (but no chance of result overflow). The result must be multiplied by the coefficients—another multiply (all coefficients are fractions only, so no overflow of the result here either)—there is an addition or subtraction of all the terms (but no chance of exceeding the maximum value, as the terms are all less than 1.0), and you have the result of the polynomial operation.

The preceding argument is presented to express the fact that we can safely compute the polynomial solutions using the fixed-point format alone, and that is the format being used to develop the library of functions in the next section.

The ideal implementation of the fixed-point format would be to use the native CPU instructions of the 32-bit integer add, subtract, multiply, and divide. However, that would require an interface with the Assembly language routines, which would cause portability problems across different platforms. Instead, the basic routines are simulated in the C language functions using the integer operations of the

C compiler. Interested readers could rewrite the functions in Assembly language to gain further performance improvement and optimize the speed even more.

A Fast and Efficient Algorithm for the Square Root Function

Another example of a fixed-point implementation is a solution of the sqrt function. The algorithm is based on the fact that a floating-point number can easily be split into its exponent and fraction portion, and each can be treated as a separate, fixed-point quantity. Think about a floating-point number in terms of its scientific notation with the exponent portion multiplied by a fraction portion. Using the multiplicative property of the square root function, finding the square of a number is simply the square root of the exponent portion multiplied with the square root of the fraction portion, as shown here:

$$\sqrt{2^x * fraction} = \sqrt{2^x} * \sqrt{fraction}$$

For example, the square root of the number 3.0 is equivalent to:

$$\begin{aligned}\sqrt{3} &= \sqrt{2^2 * 0.75} \\ &= \sqrt{2^2} * \sqrt{0.75}\end{aligned}$$

The sqrt of the exponent portion is obtained by shifting the exponent to the right 1 bit; in other words, dividing it by 2.

$$= 2^1 * \sqrt{0.75}$$

The fraction quantity is always less than 1.0 by definition, so we can apply the same fixed-point technique we developed in the previous section of transcendental functions. The sqrt of a 32-bit fraction is obtained by a binary search through maximum 16 iterations of 16-bit by 16-bit multiplication of an approximation. At the end, the individual results are simply combined into floating-point format and returned. There are two steps in this algorithm:

1. sqrt of exponent = exponent shift right 1
2. sqrt of fraction = X, if (X x X = fraction)

See lines 430 through 555 in Listing 8.1 for an implementation of the algorithm.

TRANSCENDENTAL FUNCTIONS LIBRARY

The following is a transcendental functions library of routines for the C language interface based on the fixed-point algorithm discussed in this chapter. The routines are divided into three sections:

- Basic arithmetic operation implementation
- General solution of the polynomial equation
- Exported transcendental functions

The basic arithmetic operations include fixed-point to floating-point, floating-point to fixed-point conversion, and floating-point multiply routines simulated in C. Just to give you an idea of how easy it is to write fixed-point multiply in Assembly language of Intel 80386, the following Assembly language instructions are equivalent of the library function fx_mpy() in Listing 8.1 (lines 207 through 241).

Assuming EAX has the multiplier and EDX has the multiplicand:

```
mul EDX
shrd EAX,EDX,24
```

There is a general polynomial solution in Listing 8.1, (lines 250 through 277). The three input fixed-point value parameters are:

1. Input x;
2. List of coefficients
3. Number of coefficients in the list

The polynomial return is a double value, so the mundane computations could be performed with the help of C compiler library routines.

The exported functions are:

```
double __sin(double x);
double __cos(double x);
double __tan(double x);
double __exp(double x);
double __lge(double x);
double __lgd(double x);
double __sqrt(double x);
```

The function names have been appended with the underscore (e.g., __cos(x)) to differentiate between fixed-point library functions and the original floating-point math library.

ON THE CD The source code is presented in the companion CD folder /book/chapter8/mathlib.c.

LISTING 8.1 file book/chapter8/mathlib.c fixed point arithmetic library.

```
1)   /************* Transcendental Functions ***********
2)   The sin, cos, tan, exp, ln and square root function
3)   solution are presented using fixed point arithmetic
4)   and reduced coefficient polynomials.
5)   You can use the file as an include or compile into
6)   a library for linking with other modules. The names
7)   of the functions are appended with an underscore such
8)   as __cos so that it does not conflict with the default
9)   library routines.
10)  The followings are the exported functions
11)  double __sin(double x);
12)  double __cos(double x);
13)  double __tan(double x);
14)  double __exp(double x);
15)  double __lge(double x);
16)  double __lgd(double x);
17)  double __sqrt(double x);
18)  **************************************************/
19)  #include <stdio.h>
20)  #include <math.h>
21)  #define FLOATDIGITS    6
22)  #define MAXINTEGER 2.147483647e+9
23)  #define debug
24)  #define test
25)  #define fixed
26)  #define TWO_POINT_ZERO 2.0
27)  #define TWO_PI_CONST 6.283185
28)  #define PI_CONST 3.141593
29)  #define PI_2_CONST 1.570796
30)  #define PI_4_CONST 0.7853982
31)  #define ONE_POINT_ZERO 1.0
32)  #define CONST_POINT_693 0.693
33)  #define CONST_LN_2 0.69314718
34)  #define CONST_LN_10 2.302585093
35)  #define TWO_POINT_ZERO 2.0
36)  #ifndef INFINITY
37)  #define INFINITY 1.0E99
38)  #endif
```

```
39)  #ifndef NAN
40)  #define NAN 0.5E99
41)  #endif
42)  #define modulus(r1,r2) (r1- (((int) (r1/r2))* r2))
43)  /* float xtemp,ytemp; */
44)  /********************************************
45)  The coeffiecients of the polynomials are converted
46)  into fixed point format as the polynomial solution
47)  requires the parameters to be presented in a fixed-
48)  point format.
49)  ********************************************/
50)  #ifdef fixed
51)  long SIN_COEF_TBL[5] = {0x1000000,0,0xffd57dc0,0,0x1f2ba};
52)  long COS_COEF_TBL[5] = {0x1000000,0,0xff80d845,0,0x97c1b};
53)  long COT_COEF_TBL[5] = {0x1000000,0,0xffaac93b,0,0xfff9c2f5};
54)  long ATAN_COEF_TBL[10] = {0,0xfff738,0,0xffab717e,0,0x2e1db8,0,0xffea34ba,
55)  0,0x55572};
56)  long LN_COEF_TBL[6] = {0,0xffdef1,0xff821241,0x4a1b05,0xffdd2afe,0x83b89};
57)  long EXP_COEF_TBL[5] = {0x1000000,0xff0008a0,0x7f901a,0xffd728d5,0x78467};
58)  #else
59)  float SIN_COEF_TBL[5] = {1.0,0.0,-0.16605,0.0,0.00761};
60)  float COS_COEF_TBL[5] = {1.0,0.0,-0.49670,0.0,0.03705};
61)  float COT_COEF_TBL[5] = {1.0,0.0,-0.332867,0.0,-0.024369};
62)  float ATAN_COEF_TBL[10] = {0.0,0.9998660,0.0,-0.3302995,0.0,
63)  0.1801410,0.0,-0.0851330,0.0,0.0208351};
64)  float LN_COEF_TBL[6] = {0.0,0.99949556,-0.49190896,0.28947478,
65)  -0.13606275,0.03215845};
66)  float EXP_COEF_TBL[5] = {1.0,-0.9998684,0.4982926,-0.1595332,0.0293641};
67)  #endif
68)  /*********************************************************
69)  Exponent is calculated as exponent of integer + exponent
70)  of fraction. Exponent of up to 32 integers are sufficient
71)  for all practical purpose, beyond that it might as well
72)  be infinity.
73)  A lookup table of exponent of integers is presented below.
74)  *********************************************************/
75)  float INT_EXP_TBL[33] = {
76)  1.000000000,
77)  2.718281828,
78)  7.389056099,
79)  20.08553692,
80)  54.59815003,
81)  148.4131591,
82)  403.4287935,
```

```
83)  1096.633158,
84)  2980.957987,
85)  8103.083927,
86)  22026.46580,
87)  59874.14172,
88)  162754.7914,
89)  442413.3920,
90)  1202604.284,
91)  3269017.373,
92)  8886110.521,
93)  24154952.75,
94)  65659969.14,
95)  178482301.0,
96)  485165195.4,
97)  1318815735.0,
98)  3584912846.0,
99)  9744803446.0,
100) 2.6489122E10,
101) 7.2004899E10,
102) 1.9572960E11,
103) 5.3204824E11,
104) 1.4462570E12,
105) 3.9313342E12,
106) 1.0686474E13,
107) 2.9048849E13,
108) 1.0E38};
109) /****************** function define ****************/
110) double __sin();
111) double __cos();
112) double __tan();
113) double __exp();
114) double __lge();
115) double __lgd();
116) double __atan();
117) double fx2fl();
118) long fl2fx();
119) long fx_mpy();
120) /*****************************************************
121) general purpose conversion routines
122) *****************************************************/
123) /***************floating point to fixed point *********/
124) long    fl2fx(double x)
125) {
126) long *k,mantissa,exp;
```

```
127) float temp;
128) temp = (float) x;
129) /* get a pointer to fl pt number */
130) k = (long *) & temp;
131) /* remove sign bit from the number */
132) mantissa =(*k & 0x7fffffff);
133) /* extract exponent from the mantissa */
134) exp = ((mantissa >> 23) & 0xff);
135) if (!exp)
136) return (0);
137) /* mantissa portion with normalized bit */
138) mantissa = ((mantissa & 0xffffff) | 0x800000);
139) /* inxrease exp for bringing in normalize bit */
140) exp++;
141) if (exp > 0x7f)
142) {
143) /* exp > 0x7f indicates integer portion exist */
144) exp = exp - 0x7f;
145) mantissa = mantissa << exp;
146) }
147) else
148) {
149) /* exp <= 0x7f indicates no integer portion */
150) exp = 0x7f - exp;
151) mantissa = mantissa >> exp;
152) }
153) if (*k >= 0)
154) return(mantissa);
155) else
156) return (-mantissa);
157) }
158) /************* fixed point to floating point *********/
159) double fx2fl(long j)
160) {
161) long i,exp;
162) float temp,*fp;
163) i = j;
164) if (i == 0)
165) return ((double)(i));
166) /*
167) maximum exp value for an 8-bit int and 24-bit fraction
168) fixed point
169) biasing factor = 7f;
170) 7f + 8 = 87;
```

```
171) */
172) exp = 0x87;
173) if (i < 0)
174) {
175) exp = exp | 0x187;
176) i = -i;
177) }
178) /* normalize the mantissa */
179) do
180) {
181) i = i << 1; exp –;
182) }
183) while (i > 0);
184) i = i << 1; exp –;
185) /* place mantissa on bit-0 to bit-23 */
186) i = ((i >> 9) & 0x7fffff);
187) exp = exp << 23;
188) i = i | exp;
189) fp = (float *) & i;
190) temp = *fp;
191) return ((double) temp);
192) }
193) /*************** fixed point multiply ****************/
194) long fx_mpy(long x,long y)
195) {
196) unsigned long xlo,xhi,ylo,yhi,a,b,c,d;
197) long sign;
198) /* The result sign is + if both sign are same */
199) sign = x ^ y;
200) if (x < 0)
201) x = -x;
202) if (y < 0)
203) y = -y;
204) /*
205) two 32-bit numbers are multiplied as
206) if there are four 16-bit words
207) */
208) xlo = x & 0xffff;
209) xhi = x >> 16;
210) ylo = y & 0xffff;
211) yhi = y >> 16;
212) a = xlo * ylo;
213) b = xhi * ylo;
214) c = xlo * yhi;
```

```
215) d = xhi * yhi;
216) a = a >> 16;
217) /* add all partial results */
218) a = a+b+c;
219) a = a >> 8;
220) d = d << 8; a = a+d;
221) if (sign < 0)
222) a = -a;
223) return(a);
224) }
225) /*************** fixed point polynomial ****************/
226) /*
227) Three input parameters; all fixed point
        1-      input x;
        2-      list of coefficients
        3-      number of coefficients in the list
228) */
229) double polynomial_x(double r1,long * r2,int i)
230) {
231) long j,k,temp,sum;
232) k = fl2fx(r1);
233) sum = 0;
234) /* steps
        1-      find power of x,
        2-      multiply with coefficient
        3-      add all terms
235) */
236) while(i—)
237) {
238) temp = 0x1000000; /* this is 1.0 */
239) if (r2[i] != 0)
240) {
241) /* power of x */
242) for (j=i;j>0;j—)
243) temp = fx_mpy(temp,k);
244) /* and then multiply with coefficient */
245) temp = fx_mpy(temp,r2[i]);
246) }
247) else
248) temp = 0;
249) sum = sum + temp;
250) };
251) return(fx2fl(sum));
252) }
```

```
253) /********************** sin ************************/
254) double __sin(double x)
255) {
256) float temp;
257) int quadrent;
258) unsigned long *k;
259) temp = (float) x;
260) /* make absolute value */
261) k = (unsigned long *) &temp;
262) *k = (*k & 0x7fffffff);
263) /* 360 degree modulus */
264) if (temp > TWO_PI_CONST)
265) {
266) temp = modulous (temp,TWO_PI_CONST);
267) }
268) /* negative angles are made complements of 360 degree */
269) if ( x < 0)
270) temp = TWO_PI_CONST - temp;
271) quadrent = (int) (temp / PI_2_CONST);
272) temp = modulus(temp,PI_CONST);
273) if (temp > PI_2_CONST)
274) temp = (PI_CONST - temp);
275) temp = (polynomial_x(temp,SIN_COEF_TBL,5) * temp);
276) if (quadrant >= 2)
277) temp = -temp;
278) return (temp);
279) }
280) /********************** cos ************************/
281) double __cos(double x)
282) {
283) float temp;
284) int quadrent;
285) unsigned long *k;
286) temp = (float) x;
287) /* make absolute value */
288) k = (unsigned long *) &temp; *k = (*k & 0x7fffffff);
289) /* 360 degree modulus */
290) if (temp > TWO_PI_CONST)
291) {
292) temp = modulus (temp,TWO_PI_CONST);
293) }
294) /* negative angles are made complements of 360 degree */
295) if ( x < 0)
296) temp = TWO_PI_CONST - temp;
```

```
297) /* find out the quadrant from the original angle */
298) quadrant = (int) (temp / PI_2_CONST);
299) temp = modulus(temp,PI_CONST);
300) if (temp > PI_2_CONST)
301) temp = (PI_CONST - temp);
302) temp = (PI_2_CONST - temp );
303) temp = (polynomial_x(temp,SIN_COEF_TBL,5) * temp);
304) if ((quadrant ==1) || (quadrant == 2)) temp = -temp;
305) return ((float) (temp));
306) }
307) /********************* tan ***********************/
308) double __tan(double x)
309) {
310) float temp;
311) int quadrant;
312) unsigned long *k;
313) temp = (float) x;
314) /* make absolute value */
315) k = (unsigned long *) &temp;
316) *k = (*k & 0x7fffffff);
317) /* 360 degree modulus */
318) if (temp > TWO_PI_CONST)
319) {
320) temp = modulus (temp,TWO_PI_CONST);
321) }
322) /* negative angles are made complements of 360 degree */
323) if ( x < 0)
324) temp = TWO_PI_CONST - temp;
325) quadrant = (int) (temp / PI_2_CONST);
326) temp = modulus(temp,PI_CONST);
327) if (temp > PI_2_CONST)
328) temp = (PI_CONST - temp);
329) if (temp < PI_4_CONST)
330) {
331) temp=((ONE_POINT_ZERO /
332) polynomial_x(temp,COT_COEF_TBL,5)) * temp);
333) }
334) else
335) {
336) temp = PI_2_CONST - temp;
337) temp = ONE_POINT_ZERO /
338) ((ONE_POINT_ZERO /polynomial_x(temp,COT_COEF_TBL,5)) * temp);
339) }
340) if ((quadrant == 1)||(quadrant==3))
```

```
341) temp = -temp;
342) return (temp);
343) }
344) /********************* exp ***********************/
345) /* e pwr x = 2 pwr (x * log2_e) */
346) /* if |X| <= 0.693.. then exp(x) = exp(fraction) */
347) /* else exp(x) = exp(INT + 0.693 + FRACTION) */
348) /* exp(0.693) = 2;*/
349) double __exp(double x)
350) {
351) float temp = x;
352) int i;
353) unsigned long *k;
354) if (x > 0.0)
355) i= ((int) (x + 0.0000001));
356) else
357) i= ((int) (x - 0.0000001));
358) if (i < 0)
359) i = 0-i;
360) if (i > 31)
361) {
362) if (x > 0)
363) return(INT_EXP_TBL[32]);
364) else
365) return(0);
366) }
367) temp = (float) x;
368) /* make absolute value */
369) k = (unsigned long *) &temp;
370) *k = (*k & 0x7fffffff);
371) temp = modulus(temp,ONE_POINT_ZERO);
372) if (temp > CONST_POINT_693)
373) {
374) temp = temp - CONST_POINT_693;
375) temp = ONE_POINT_ZERO /
376) (polynomial_x(temp,EXP_COEF_TBL,5));
377) temp = temp * TWO_POINT_ZERO;
378) }
379) else
380) {
381) temp = ONE_POINT_ZERO /
382) (polynomial_x(temp,EXP_COEF_TBL,5));
383) }
384) temp = (INT_EXP_TBL[i]) * temp;
```

```
385) if (x < 0)
386) temp = ONE_POINT_ZERO / temp;
387) return (temp);
388) }
389) /********************* ln (x)***********************/
390) /* log base E (X) = ln (x)/ln (E) */
391) /* ln (x) = pwr * ln(2) + ln(fraction+1;*/
392) /* range =(0 <= x <= 1)*/
393) double  __lge(double x)
394) {
395) static long exp,i,*k;
396) static float temp,xtemp,*fraction;
397) xtemp = (float) x;
398) temp=0;
399) k =(long *) &xtemp;    i=k[0];
400) /* do not compute ln of a minus number */
401) if (i <= 0)
402) return (NAN); /* NAN could be replaced by 0.0 */
403) /*remove exp and sign and check for fraction*/
404) if ((i =(i & 0x7fffff)) != 0)
405) {
406) exp = 0x7f;
407) while ((i & 0x800000) == 0)
408) {
409) i = i << 1;  exp—;
410) }
411) i = i & 0x7fffff;         /* remove normalized bit */
412) exp = exp <<23;     /* new exp for fraction */
413) i = i | exp;        /* combine exp and mantissa */
414) fraction = (float *) &i;
415) temp = fraction[0];
416) temp = (polynomial_x(temp,LN_COEF_TBL,6));
417) }
418) /* from input value x extract the exp */
419) exp = *k;
420) exp = exp >> 23;
421) exp = exp - 0x7f;
422) temp = temp + (exp * CONST_LN_2);
423) return(temp);
424) }
425) /********************* lgd ***********************/
426) double __lgd(double x)
427) {
428) return (__lge(x)/CONST_LN_10);
```

```
429) }
430) /********************* square root *********************
431) A fast and efficient algorithm for Square Root function.
432) The algorithm is based on the fact that a floating point number
433) can be easily split into its exponent and fraction portion and
434) each can be treated as separate integer quantity. The sqrt of
435) exponent portion is obtained by shifting the exponent right 1-
436) bit, i.e., dividing it by 2 and the sqrt of a 32-bit fraction is
437) done by a binary search of 65536 possible values or 16 trials of
438) 16-bit x16-bit multiplication of an approximation. At the end,
439) the individual results are simply combined into floating point
440) format and returned.
441) sqrt of exponent  = exponent shift right 1    ........ (1)
442) sqrt of fraction  = X, if (X x X = fraction)  ........ (2)
443) *******************************************************/
444) double __sqrt(double x)
445) {
446) float xtemp;
447) unsigned long exp,i,j,*k;
448) unsigned short int new_apprx,old_apprx;
449) xtemp = (float) x;
450) /*
451) step 1—— Think of the floating point number as an
       1-     integer quantity
452) */
453) k =(unsigned long *) &xtemp;    i=k[0];
454) /*
455) step 2—— do not compute sqrt of a minus number
456) */
457) if (x <= 0.0)
458) {
459) if (x==0.0)
460) return(x);
461) else
462) return (NAN); /* NAN could be replaced by 0.0 */
463) }
464) /*
465) step 3—- extract exp and place it in a byte location
466) */
467) exp = (i & 0x7f800000);
468) exp >>= 23;
469) /*
470) step 4 —- bring fraction portion to 32-bit position and
       1-     bring normalized bit
```

```
471)  */
472)  i <<= 8;
473)  i = i | 0x80000000;
474)  /*
475)  step 5 –- inc exp for bringing in normal bit and
        1-     remove exp bias
476)  */
477)  exp += 1;
478)  exp -= 0x7f;
479)  /*
480)  step 6 –- compute square root of exp by shift
        1-     right or divide by 2
        2-     if exp is odd then shift right mantissa
        3-     one more time
481)  */
482)  if ( exp & 1)
483)  {
484)  i >>= 1;
485)  exp >>= 1;
486)  exp += 1;
487)  }
488)  else
489)  exp >>= 1;
490)  /*
491)  step 7–- add bias to exp
492)  */
493)  exp += 0x7f;
494)  /*————-sqrt of fraction ————-*/
495)  /*
496)  step 8 —-start with 0x8000 as the first apprx
497)  */
498)  j = old_apprx = new_apprx = 0x8000;
499)  /*
500)  step 9 –-use binary search to find the match.
        1-     loop count maximum 16
501)  */
502)  do
503)  {
504)  /*
505)  step 9a — multiply approximate value to itself
506)  */
507)  j = j*j;
508)  /*
```

```
509)  step 9b — next value to be added in the binary
        i.     search
510)  */
511)  old_apprx >>= 1;
512)  /*
513)  step 9c — terminate loop if match found
514)  */
515)  if (i == j )
516)  old_apprx = 0;
517)  else
518)  {
519)  /*
520)  step 9d -- if approx. exceeded the real fraction
        i.     then  lower the approximate
521)  */
522)  if (j > i )
523)  new_apprx -= old_apprx;
524)  /*
525)  step 9e -- else increase the the approximate
526)  */
527)  else
528)  new_apprx += old_apprx;
529)  }
530)  j =new_apprx;
531)  } while (old_apprx != 0);
532)  /*
533)  step 10 --  combine exp and mantissa
534)  */
535)  j =new_apprx;
536)  /*
537)  step 10a -- bring mantissa to bit position 0..23
        i.     and remove Normal bit
538)  */
539)  j <<= 8;
540)  j &= 0x7fffff;
541)  /*
542)  step 10b -- decrement exponent for removing Normalized
        i.     bit
543)  */
544)  exp -= 1;
545)  /*
546)  step 10c -- bring exponent to position 23..30
547)  */
548)  exp <<= 23;
```

```
549) /*
550) step 10d –- combine exponent and mantissa and
     i.      return double
551) */
552) j = j | exp;
553) xtemp = * ((float *) &j);
554) return (xtemp);
555) }
```

Testing the Transcendental Function Library

A comprehensive confidence test is being prepared (Listing 8.2) for verification of the routines provided in the mathlib.c library of Listing 8.1.

Test Objectives

The objective of this test is to verify the accuracy of the fixed-point transcendental function library routines.

Test Description

This will test the sin, cos, tan, ln, exp, and square root functions for a specified range of numbers. The test compares the result from the fixed-point library vs. the default floating-point library and prints the magnitude difference between the computed values and the published values. A test program is shown in Listing 8.2. Each test case comprises a range of numbers from extreme negative values to extreme positive values.

ON THE CD The source code is presented in the companion CD folder /book/chapter8/test.c.

LISTING 8.2 File book/chapter8/test.c test program for verification math library routines.

```
1)    /***************************************************/
2)    /*
      •    calculating error magnitude between fixed-point result
      •    and floating-point original math library result

3)    */
4)    #include "mathlib.c"
5)    #include <math.h>
6)    static double greatest_err = 0.00001;
7)    /***************************************************/
8)    static double err_magnitude(double calculated, double original)
```

```
9)   {
10)  double err;
11)  err = (original - calculated);
12)  if (err < 0.0)
13)  err = -err;
14)  if (err < 0.0001)
15)  return 0.0001;
16)  if (err > greatest_err)
17)  greatest_err = err;
18)  return err;
19)  }
20)  /************** test case routine *****************/
21)  /*
22)  Compare the fixed-point transcendental functions
23)  with the floating-point library functions
24)  The routine accepts the following parameter
          1-      beginning value
          2-      ending value
          3-      increment
25)  */
26)  static int test_case(double begin, double end,double  inc)
27)  {
28)  double x , original, calculated;
29)  double difference;
30)  greatest_err = 0.00001;
31)  for (x= begin ; x < end;  x+= inc)
32)  {
33)  if (x < 15)
34)  {
35)  original = exp(x);
36)  calculated = __exp(x);
37)  difference = err_magnitude(calculated, original);
38)  }
39)  }
40)  printf (" EXP: %f < x < %f --, Max Difference <=
          1-      %f\n",begin,end, greatest_err);
41)  greatest_err = 0.00001;
42)  for (x= begin ; x < end;  x+= inc)
43)  {
44)  original = sin(x);
45)  calculated = __sin(x);
46)  difference = err_magnitude(calculated, original);
47)  }
```

```
48)  printf (" SIN: %f < x < %f --, Max Difference <=
          1-      %f\n",begin,end, greatest_err);
49)  greatest_err = 0.00001;
50)  for (x= begin ; x < end;  x+= inc)
51)  {
52)  original = cos(x);
53)  calculated = __cos(x);
54)  difference = err_magnitude(calculated, original);
55)  }
56)  printf (" COS: %f < x < %f --, Max Difference <=
          1-      %f\n",begin,end, greatest_err);
57)  greatest_err = 0.00001;
58)  for (x= begin ; x < end;  x+= inc)
59)  {
60)  original = tan(x);
61)  calculated = __tan(x);
62)  difference = err_magnitude(calculated, original);
63)  }
64)  printf (" TAN: %f < x < %f --, Max Difference <=
          1-      %f\n",begin,end, greatest_err);
65)  greatest_err = 0.00001;
66)  for (x= begin ; x < end;  x+= inc)
67)  {
68)  if(x > 0.00001 )
69)  {
70)  original = log(x);
71)  calculated = __lge(x);
72)  difference = err_magnitude(calculated, original);
73)  }
74)  }
75)  if(x > 0.00001 )
76)  {
77)  printf (" LOG: %f < x < %f --, Max Difference <=
          1-      %f\n",begin,end, greatest_err);
78)  }
79)  greatest_err = 0.00001;
80)  for (x= begin ; x < end;  x+= inc)
81)  {
82)  if(x > 0.0 )
83)  {
84)  original = sqrt(x);
85)  calculated = __sqrt(x);
86)  difference = err_magnitude(calculated, original);
87)  }
```

```
88)  }
89)  printf (" SQRT: %f < x < %f --, Max Difference <=
         1-     %f\n",begin,end, greatest_err);
90)  return 0;
91)  }
92)  /*****************************************************/
93)  /************* main test program *********************/
94)  int main()
95)  {
96)  test_case(-1.0, 1.0, .1);
97)  test_case(-10.0, 10.0, 0.01);
98)  test_case(-100.0, 100.0, .1);
99)  test_case(-1000.0, 1000.0, 1);
100) test_case(-10000.0, 10000.0, 10.1);
101) test_case(-100000.0, 100000.0, 100.1);
102) test_case(-1000000.0, 1000000.0, 1000.1);
103) test_case(-10000000.0, 10000000.0, 10000.1);
104) test_case(-100000000.0, 100000000.0, 100000.1);
105) return (0);
106) }
```

Test Inputs

ON THE CD

1. Mount the CD-ROM if you are planning to copy the source code from the CD.

```
[root@ns /root]# mount -t iso9660 /de/cdrom /mnt/cdrom
```

ON THE CD

2. Create a subdirectory /book/chapter8/ and copy the test program and the mathlib.c program from the CD-ROM directory /book/chapter8.

```
[root@ns /root]# mkdir /book/chapter8
[root@ns/root]# cp /mnt/cdrom/book/chapter8/*/book/chapter8/
```

3. Compile and link the test.c using the make file 'Makefile'with the following make command.

```
[root@ns /root]#cd /book/chapter8; make
```

4. Execute the file.

```
[root@ns /book/chapter8]#./t Expected Test Results
```

The following is the screen dump of the results obtained from running the test. The magnitude of error in most cases is negligible except in some cases when the result value is extremely high, such as tan values close to 90 degree. It is still within the reasonable range as expected from values close to infinity

```
EXP: -1.000000 < x < 1.000000 —-, Max Difference <= 0.000406
SIN: -1.000000 < x < 1.000000 —-, Max Difference <= 0.000122
COS: -1.000000 < x < 1.000000 —-, Max Difference <= 0.000164
TAN: -1.000000 < x < 1.000000 —-, Max Difference <= 0.000010
LOG: -1.000000 < x < 1.000000 —-, Max Difference <= 0.000010
SQRT: -1.000000 < x < 1.000000 —-, Max Difference <= 0.000010
EXP: -10.000000 < x < 10.000000 —-, Max Difference <= 3.263702
SIN: -10.000000 < x < 10.000000 —-, Max Difference <= 0.000164
COS: -10.000000 < x < 10.000000 —-, Max Difference <= 0.000164
TAN: -10.000000 < x < 10.000000 —-, Max Difference <= 2.294101
LOG: -10.000000 < x < 10.000000 —-, Max Difference <= 0.000010
SQRT: -10.000000 < x < 10.000000 —-, Max Difference <= 0.000010
EXP: -100.000000 < x < 100.000000 —-, Max Difference <= 442.011182
SIN: -100.000000 < x < 100.000000 —-, Max Difference <= 0.000165
COS: -100.000000 < x < 100.000000 —-, Max Difference <= 0.000165
TAN: -100.000000 < x < 100.000000 —-, Max Difference <= 0.831879
LOG: -100.000000 < x < 100.000000 —-, Max Difference <= 0.000010
SQRT: -100.000000 < x < 100.000000 —-, Max Difference <= 0.000244
EXP: -1000.000000 < x < 1000.000000 —-, Max Difference <= 0.034165
SIN: -1000.000000 < x < 1000.000000 —-, Max Difference <= 0.000170
COS: -1000.000000 < x < 1000.000000 —-, Max Difference <= 0.000170
TAN: -1000.000000 < x < 1000.000000 —-, Max Difference <= 1.866331
LOG: -1000.000000 < x < 1000.000000 —-, Max Difference <= 0.000010
SQRT: -1000.000000 < x < 1000.000000 —-, Max Difference <= 0.000485
EXP: -10000.000000 < x < 10000.000000 —-, Max Difference <= 0.017315
SIN: -10000.000000 < x < 10000.000000 —-, Max Difference <= 0.000869
COS: -10000.000000 < x < 10000.000000 —-, Max Difference <= 0.000866
TAN: -10000.000000 < x < 10000.000000 —-, Max Difference <= 692.437081
LOG: -10000.000000 < x < 10000.000000 —-, Max Difference <= 0.000010
SQRT: -10000.000000 < x < 10000.000000 —-, Max Difference <= 0.001948
EXP: -100000.000000 < x < 100000.000000 —-, Max Difference <= 0.000010
SIN: -100000.000000 < x < 100000.000000 —-, Max Difference <= 0.007855
COS: -100000.000000 < x < 100000.000000 —-, Max Difference <= 0.007885
TAN: -100000.000000 < x < 100000.000000 —-, Max Difference <= 7376.778134
LOG: -100000.000000 < x < 100000.000000 —-, Max Difference <= 0.000010
SQRT: -100000.000000 < x < 100000.000000 —-, Max Difference <= 0.007789
```

EXP: -1000000.000000 < x < 1000000.000000 —-, Max Difference <= 0.000010
SIN: -1000000.000000 < x < 1000000.000000 —-, Max Difference <= 0.072403
COS: -1000000.000000 < x < 1000000.000000 —-, Max Difference <= 0.073354
TAN: -1000000.000000 < x < 1000000.000000 —-, Max Difference <= 26188.929802
LOG: -1000000.000000 < x < 1000000.000000 —-, Max Difference <= 0.000010
SQRT: -1000000.000000 < x < 1000000.000000 —-, Max Difference <= 0.015605
EXP: -10000000.000000 < x < 10000000.000000 —-, Max Difference <= 0.000010
SIN: -10000000.000000 < x < 10000000.000000 —-, Max Difference <= 0.927301
COS: -10000000.000000 < x < 10000000.000000 —-, Max Difference <= 0.919399
TAN: -10000000.000000 < x < 10000000.000000 —-, Max Difference <= 4243.122343
LOG: -10000000.000000 < x < 10000000.000000 —-, Max Difference <= 0.000010
SQRT: -10000000.000000 < x < 10000000.000000 —-, Max Difference <= 0.062424
EXP: -100000000.000000 < x < 100000000.000000 —-, Max Difference <= 0.000010
SIN: -100000000.000000 < x < 100000000.000000 —-, Max Difference <= 1.999787
COS: -100000000.000000 < x < 100000000.000000 —-, Max Difference <= 1.980583
TAN: -100000000.000000 < x < 100000000.000000 —-, Max Difference <= 43490.596024
LOG: -100000000.000000 < x < 100000000.000000 —-, Max Difference <= 0.000010
SQRT: -100000000.000000 < x < 100000000.000000 —-, Max Difference <= 0.248733

SUMMARY

In this chapter, we developed an efficient transcendental math emulation library suitable for low-cost embedded systems. If the emphasis is on speed rather than precision, then, in the absence of hardware math-coprocessors, the alternative is fixed-point arithmetic. There are several applications, such as graphics routines, that can take advantage of the high-speed solution offered by the library presented in this chapter. A unique solution to the square root function was also presented for floating-point numbers that is several orders of magnitude faster than a comparable library routine.

9 Embedded System Design Projects

In this chapter

- The Product Design Process
- Practical, Hands-on Design Projects
- DoorLock
- DoorLock Enhanced
- TimeIsMoney
- DrillingRig
- TankVolume

INTRODUCTION

In this chapter, we conclude the big picture of the embedded system design process and try to put together the little pieces of the puzzle. We present some fictitious projects, including a Door Lock mechanism, an Employee Entrance and Time Log system, a Platform Drilling Rate measuring system, and a Liquid Tank Level monitoring system. The projects were chosen for their simplicity of design and to illustrate that a working model can be put together using only the parallel port of a desktop PC and Linux as the operating system. Now is the time to prove the theory we have been discussing throughout the book that the embedded system design process is comprised of three components, a system interface, a user interface, and a database.

The system interface is the device driver, and the client-server mechanism and the user interface is the Web browser, the Web server, and the common gateway interface (CGI). The database is the glue that holds the system interface and the user interface together. Using the Web as the user interface opens our system to the worlds of the Internet and intranet, and fulfills our dream of carrying our product over the Internet. Having the Web as our user interface also frees up a tremendous amount of engineering work that goes on in the product design process, thus enlarging the window of opportunity every company yearns for. Let us begin by setting criteria for an embedded system design and guidelines of what not to do while designing a system.

THE PRODUCT DESIGN PROCESS

The definition of *embedded system design* depends on whom you ask. For some, it is the little micro controller hidden behind a microwave oven, or the different controls on a laundry machine. For others, it is the powerful multi-megahertz CPU-based hardware system that is printing 150 pages per minute. Whatever definition you choose, it all boils down to a CPU with some RAM, ROM, storage device, and some peripheral devices performing the desired input and output. The size is irrelevant; today's mainframe is tomorrow's hand-held calculator. So, how do you start designing an embedded system? Let us analyze the general trend in the industry and the corporate decision-making that shapes an embedded system design process.

Marketing determines the business community needs and develops preliminary specifications for the proposed product the company decides to sell. They mainly try to match price with performance, and hope to convince management that there is a need for a certain product. Then, there is scheduling and job assignments. Next, the engineering team gets the ball and the project is analyzed and the hardware is selected, usually based on current availability. There is not much choice in software selection; it is usually a C or C++ combination. A platform is chosen where the hardware and the software are supposed to live together. With such an arrangement, you can see the signs when the project is doomed to fail:

- Too big a scope for the software
- Too much interaction between the software modules
- Software modules are not clearly defined
- Communication protocols are not established
- Inflexible platform

As you can see, most of the problems are software related. "Too big a scope for the software" is the problem of asking a programmer to write code for modules that are already available; for example, writing the real-time kernel or designing a homegrown database server. (Linux is the real-time kernel, and PostgreSQL is the transaction database freely available, so why not choose them?) "Too much interaction between the software modules" is the problem of unrealistic division of the software modules. For example: Module A requests a function within Module B that really belongs to Module A. The overhead of task switching will have an adverse effect on performance. "Software modules not clearly defined"results in software that has unexpected behavior. If Module A and Module B perform similar functions but use different schemes, there is bound to be a maintenance problem. "Communication protocols are not established" is where the global variables come into play and ruin a programmer's life. Remember, modules are replaceable, but protocols are here to stay. Establish a protocol of communication between the modules and stick to the rules. "Inflexible platform"will increase the cost of the machine eventually. Hardware is not a constant; the price you pay today for certain features will probably be available tomorrow with twice the features and at half the price. Be flexible and choose off-the-shelf platform that is supported by many (make sure Linux can be ported on).

A typical project goes through the following six stages of product development cycle; you can call it the life cycle of an engineering product as it goes through the phases from concept to reality (Figure 9.1).

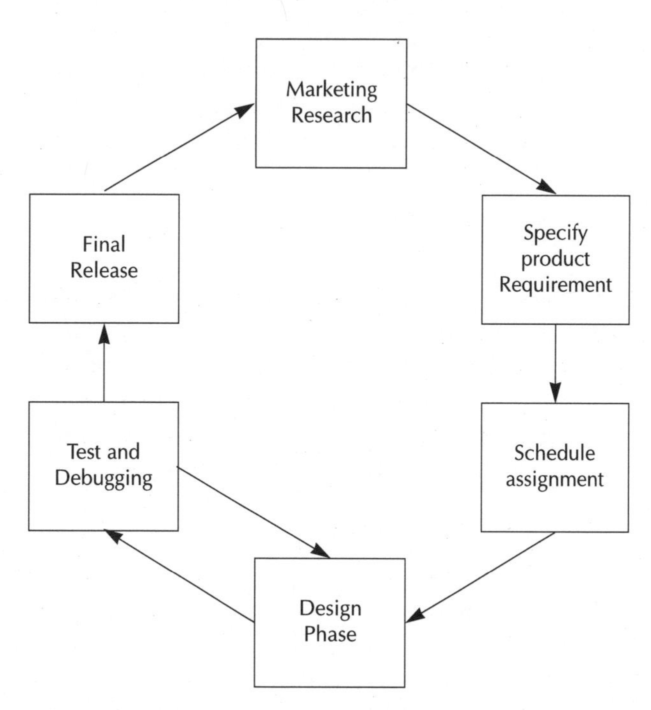

FIGURE 9.1 The six steps of the product design cycle.

1. Do the marketing research.
2. Prepare product requirement specifications.
3. Generate a development schedule.
4. Start the design process.
5. Do testing and debugging.
6. Final release.

The most critical stage in the development cycle is test and debugging. This is where you see an arrow going back to the design. You will spend most of your time between design and test and debugging. Once you get out, if you ever get out, you are home free. The best way to reduce your time in this difficult cycle is to use software modules, just as you would use prefabricated walls for constructing a big building. You will hear the term *software module* in project design meetings; everybody has his or her own definition, but the following seems most appropriate:

A software module, unless it is an independently executable code, is not truly a module at all. The classic paradigm of Input, Processing, and Output is applicable on a module only when there is a clear understanding of a client and server relationship.

It is not easy to define *what* to ask, but at least we can define *how* to ask, and that is the job of interprocess communication schemes. If you design your software around sockets and messages, you can be assured that you know whom your clients are and what questions they will ask. Acknowledging this restriction will guide you through the intricate details of establishing a client and server relationship between the modules, which will help you establish a protocol and eventually lead to a modular design.

Our Embedded System Development Philosophy

The Internet has literally changed the way we do business and the way we communicate with people. This is the age of e-mail and Web pages and the database to store the life history of every person in the world. It is no longer necessary to speak with an operator, get hold of an engineer, and request the specification of a part. Just open the browser, enter www.company.com, fill out a search form, and you will be directed to the information site right on your screen—incredible. Designing a Web system has become an assembly line job; no more complex algorithms to prove and no more costly project overruns. Perl scripting, SQL queries, and HTML pages are all you need to learn, no matter how complex the job. However, why can't we have the same success with an embedded system design? After all, a user interface is a user interface, whether it is a browser or a command line.

Embedded system design is different from Web design. In Web design, you only have a user interface and a database interface, while an embedded system de-

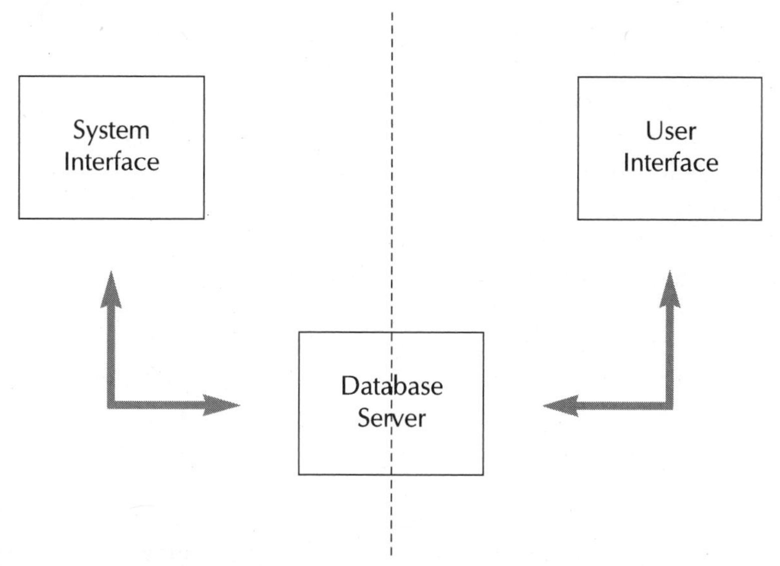

FIGURE 9.2 Embedded system design components: the system interface, the user interface, and the database server.

sign requires the added burden of the system interface (Figure 9.2). This should not stop us from using the Web concept; in fact, it can help us clearly define and isolate the user interface from the system interface—just use a database in the middle. Anything the system wants to give to the user is put in the database, and anything the user wants the system to do passes through the database.

This is a natural boundary of the functional division. No more eye contact between the user and the system; users complain too much. It is about time that embedded system designers start using transaction servers. The database design with the SQL server is probably the only mature item in the software industry that does not have a steep learning curve. Everything else, including C, C++, Java, and Visual Basic require you to have a sleeping bag right next to the computer. Application engineers, on the other hand, are quietly taking advantage of the well-defined interface between the database servers, CGI scripting, Web servers, and the Web browser.

Browsers (Netscape Navigator and Internet Explorer) are now the preferred method of displaying whatever you want the world to see, and Apache, the king of Web servers, is the engine used by more than half of Web hosting sites. The Perl Interpreter is replacing Visual Basic, and SQL will probably replace the logical part of the program design. Linux offers you all of these for free. There is PostgreSQL, the transaction server; Apache, the HTTP server; bind, the DNS server; countless application servers that universities all over the world have developed in the past several years; and, of course, the first-class real-time kernel Linux. Most of all, we are thankful to the GNU project team, whose contribution probably outweighs any other contribution.

System engineers have achieved some degree of independence in the form of device drivers. This is one piece of code that is truly a separate module. Open, close, read, and write are the only functions you have to expose if you have a device driver to interface with. Having the interface to the physical hardware set aside using a device driver, you do not have much luck with the rest of the system. The project is at the mercy of the programmer's mood. There is no way to stop a programmer from writing bad code, and programmers are wary of people who look over their shoulders. What we need to build is a client/server relationship between modules that emphasizes the use of interprocess communication mechanisms such as sockets, message queues, and threads. These are the proven mechanisms of connecting the modules together.

Unfortunately, software engineering is not an exact science. There is no set formula for an algorithm design; there should be, but we are far away from that reality. In the meantime, we need to get away from the traditional way of designing software. There is a general complaint that engineering projects involving software are usually behind schedule. The solution lies partly in building a client/server relationship between modules, and partly in integrating the Web into your product design process. By combining the best of both Web design and system design, we can extend the concept of modularity further as shown in Figure 9.3.

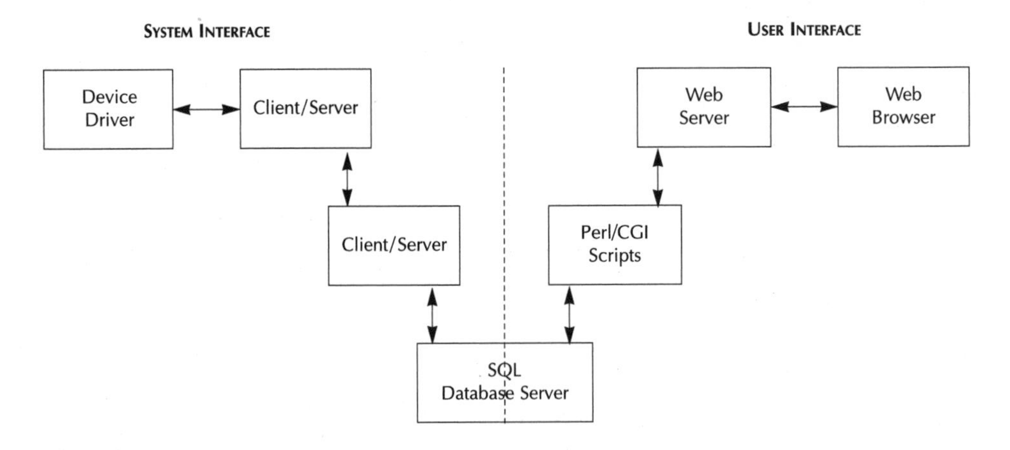

FIGURE 9.3 The interconnection of the system interface and the user interface.

The conclusion: An embedded system has a system interface and a user interface. The system interface is made up of a device driver to access the hardware, a client/server mechanism to communicate with the device driver, process the data, and eventually save the data in the database. A user interface, on the other hand, picks up the data from the database, makes an interactive Web page out of it with the help of a Perl or C script, and passes it along to a Web server, which communicates with the user and presents the page on demand. The process is reversed when a user decides to change some parameters: a Web form is filled out, the form is returned to the server, and the database is updated. The system interface picks up the new parameter from the database and makes appropriate system changes. With this simple scheme, we proceed with the challenge of designing products that incorporate the concept we just developed.

PRACTICAL HANDS-ON EXPERIENCE

Nothing works better than proving the concept practically and having the hands-on experience of designing some real-life embedded systems. If you are using PC-compatible hardware for your learning experience, then you already have a parallel port built in to the system. It is ideally suited for some quick hands-on projects. Designing an embedded system is more fun than designing an application. With embedded systems, you have total control over the entire machine and more; you are reaching out and controlling the world. Our device driver design in Chapter 3 provided easy access to the different input and output registers of the parallel port. We will use that device driver to design a few projects around the parallel port registers; it is the closest thing we have to real-life embedded design hardware.

Although the parallel port can be expanded to infinite, that would require extra circuitry. For the sake of simplicity, we will use the five input pins to simulate touch buttons as if they are the keypad switches. You can use a piece of wire connected to a ground pin to simulate the switch closing action. The interrupt pin can be used as a trigger for pulse monitoring, and if you want, you can use the output pins to set control logic. Please see the schematics in Chapter 2 for the parallel port register's pin assignments.

Now, let us explore some design ideas.

Door Lock System with Coded Key and Time Log

We will use the parallel port of the PC as a control system for a door-lock mechanism. The coded-key door-lock system logs the timing information of the last 30 entries in a database and grants access based on a coded key. The idea behind this is to demonstrate a minimal Linux implementation that cannot afford a full-blown

Web server or database server. Instead, we will use a homegrown Web server (you saw an implementation in Chapter 4) and the miniature database DBM (available in Linux for free, see Appendix B, "Miscellaneous Linux Setup"), replacing PostgreSQL.

Enhanced Door Lock System with Time-Dependent Coded Key and Time Log

This design is similar to the door-lock system discussed previously, but with added functionality such as an enhanced database and a Web interface. The coded-key enhanced door-lock system logs the timing information of the last 10,000 entries in a database, and grants access based on a coded key and time-of-day restrictions.

In this product design, a full Linux implementation is demonstrated, using PostgreSQL as the database server and a Web interface through the DNS/HTTP server.

Platform Drilling Rate Monitor with Remote Log

In this project, a process control system is being designed that interfaces with the interrupt pin and the digital I/O section of the PC parallel port. The interrupt signal from the PC parallel port is used as a rate counter and rate meter of a simulated oil-platform drilling operation.

The channel activity could be monitored over the Web.

Tank-Level Monitoring System

This project is a data acquisition system design for monitoring the liquid level of storage tanks, such as petroleum refineries, desalination plants, and so forth. The liquid-level sensor input is simulated with the input pins of the parallel port, similar to contact switches at different heights of the liquid level. The Web implementation demonstrates a live animation with the help of Perl scripts.

Employee Time and Attendance

This project is a Web-based application that can be launched through a browser for logging the employee time-in and time-out information, integrated with an entrance security system with code-key access. A supervisor can view employee attendance via a supervisor login ID.

In keeping with the ideals of our design philosophy, we will have:

The system interface:

- A device driver to acquire the external signals and control the output
- A server mechanism to communicate with the device driver
- A client mechanism to request the data from the server and store it in the database

The user interface:

- Perl/CGI scripts to acquire data from the database and create a Web page
- A Web server to transfer the Web page
- A browser to view the data over the Internet

The projects are developed as working models to prove the concept of using Internet/intranet as the basis for the user interface and device driver, and a client/server mechanism as the basis for the system interface, albeit with minimum specifications. The intent here is to put to use the material we developed in earlier chapters.

The design aspects of the projects will be covered from the point of view of a full-scale Linux implementation, where an SQL server, DNS server, and HTTP server are in place, as well as a minimal Linux implementation where the SQL server is replaced with the DBM (Database Manager), and the DNS and HTTP servers are replaced by a simple Socket server that we developed in Chapter 4 as a replacement for the Apache Web server.

PROJECT 1: SECURE YOUR ENTRANCE WITH DOORLOCK

A computerized controlled door-lock system with coded key.

Product Specifications

- A low-cost coded-key door-lock system with time log facility
- Ideal for hotel rooms with centralized control
- Time log database available through a Web interface
- Web-based viewing of time log
- Modular unit with built-in touch keypad and display

Introduction

DoorLock is a low-cost security system with a coded key and time log facility, ideal for a hotel-room locking system. The product will be developed using the Linux operating system, with the PC parallel port providing the necessary hardware circuitry.

The purpose of this implementation is to demonstrate the capability of the Linux operating system as a true low-cost embedded system, without the need for a SQL, DNS, or HTTP server.

Design Goals

An electronic door-lock security system with coded key entrance control using the guidelines established earlier in the chapter.

The supervisory user control is provided by a Web interface using a minimum function Web server. The client system interface, such as accessing a key matrix, is controlled by a device driver and a client/server mechanism using the Linux real-time kernel.

System Requirements

- Five-digit numeric keypad, simulated with the input pins of the status register of the PC parallel port
- Green LED signaling a valid key combination, controlled by the PC parallel data register bit 0
- Red LED signaling an invalid key combination, controlled by the PC parallel data register bit 1
- Lock solenoid latching signal, controlled by the PC parallel data register bit 2
- Capability of storing a time log of the last 100 entries
- Viewing capability via a browser (Netscape Navigator)

The Hardware Functional Description

The PC parallel port as defined in Chapter 3, Figure 3.1 will be used as the basic hardware component for providing the external user interface. The Linux box will be used as the CPU. The data register of the parallel port is used as the controlling signal to turn the entrance solenoid on and off, as well as the red and green status LED. The five-digit touch keypad will be simulated with the status register of the parallel port.

The Software Functional Description

The concept of embedded system design developed earlier in the chapter is used as the foundation for building the different software modules of the system. First, we have the division of the user interface and the system interface bound only by the database server. The user interface is a Web-based implementation where the supervisor can query the database entries through a browser such as Netscape being served by the mini Web server specifically designed for this low-cost implementation.

The Web server acquires the logging information from the database and creates dynamic Web pages to be displayed by the browsers. The system interface incorporates the device driver to communicate with the hardware, and a client server module that receives the coded key sequence from the device driver, verifies the entry and updates the database with the time-log information.

Functional Overview

The software modules for DoorLock are comprised of the system interface and the user interface. Figure 9.4 shows the functional block diagram and the interconnection of the software modules used in the system.

The system interface modules:

- DeviceDriver
- ClientServer

The user interface modules:

- WebServer
- DBM

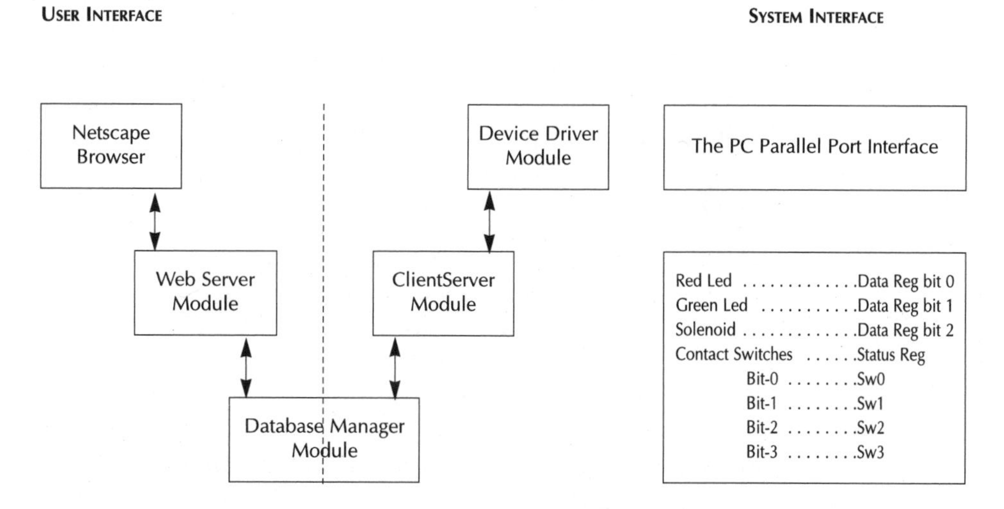

FIGURE 9.4 Block diagram showing interconnection of software modules.

The Device Driver Module

The module communicates with the hardware of the system as a standard Linux device driver. The touch keys are simulated through the status register of the parallel port. The lock opening and the output LEDs are controlled with the logic levels of the data register output signals: data bit 0 for the red LED control, data bit 1 for the green LED control ,and data bit 2 for the solenoid activation.

The following function calls are supported:

int open(void)	Provides the initialization of the hardware
int close(void)	Provides the closing of the file descriptor
int read(void)	Provides the keypad access
int write(int pattern)	Controls the logic level of the output pins

Communication Protocol through the read() Function

1. Return –1 indicating no key pressed.
2. Return 0 for closing action of SW0.
3. Return 1 for closing action of SW1.
4. Return 2 for closing action of SW2.
5. Return 3 for closing action of SW3.
6. Return 4 for closing action of SW4.

Communication Protocol through the write() Function

1. The input parameter pattern bit 0 controls the red LED logic level.
2. The input parameter pattern bit 1 controls the green LED logic level.
3. The input parameter pattern bit 2 controls the solenoid logic level.

The ClientServer Module

The module is responsible for providing the desired system interface as a client/server model. The primary responsibility of this module is to verify the user input and update the time log information in the database. The client task periodically scans the keypad, and once a valid key sequence is received, the server task is provided the message of the user key entry. Figure 9.5 is a flowchart for the client task. The server task waits for the message from the client task. When a key sequence message is received, the values are compared against the valid key values from the database. If a match is found, the server issues a door-opening sequence and logs the time of event with the database. Figure 9.6 is a flowchart for the server task module.

The Communication Protocol through CS Message Queues

The basic message ID is the USER_KEY_SEQUENCE that indicates that the contents of the message are available in the message queue.

Client Task

The client task scans the device driver for the user input, and a message is created for the server task when a sequence of key presses are received. The message ID is USER_KEY_SEQUENCE, and the message body contains the input string.

```
#define USER_KEY_SEQUENCE 1
```

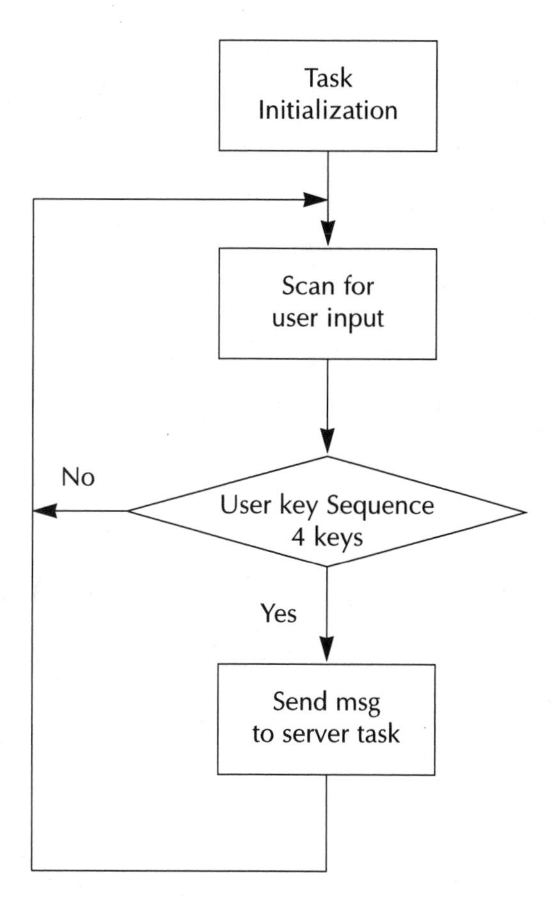

FIGURE 9.5 Flow chart for the client task.

Server Task

The server task waits for the message queue ID USER_KEY_SEQUENCE, while the body of the message has the key sequence in an ASCII string. The key sequence is verified against entries in the database, and if a match is found, the device driver write function is called with the proper gate-opening bit pattern. The current time is logged in the database.

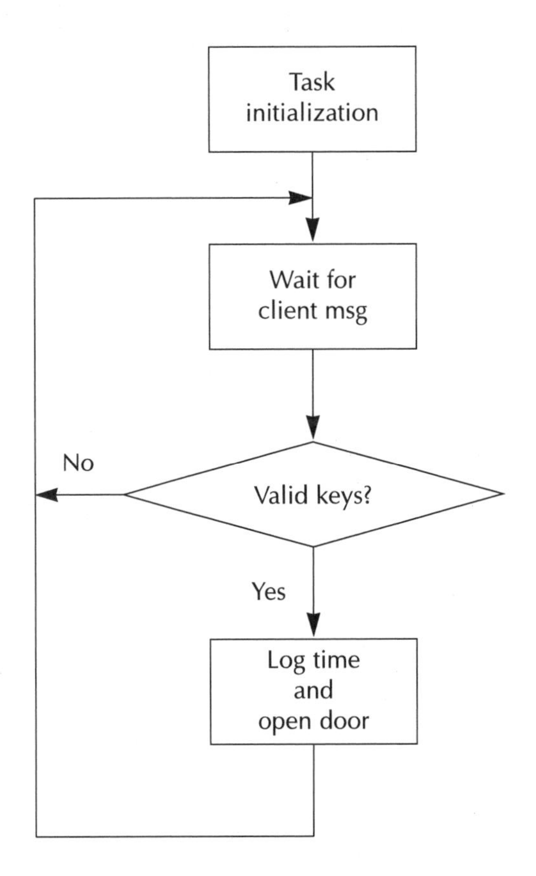

FIGURE 9.6 Flow chart for the server task.

The Web Server Module

The Web server module waits for a client connection at TCP/IP socket port 9999. Once a connection is received, a Web page is created that returns the time log information from the database.

Implementation

The module files are provided on the companion CD directory /book/chapter9/DoorLock, and the contents should be copied to a local directory for compilation and testing purposes. Detailed instructions are also available in the ReadMe.txt file (CD directory path /book/chapter9/readme.txt) section "DoorLock."

NOTE

You must be root user in order to perform the following operation.

Insert the companion CD into the CD-ROM, and mount the driver using the following command:

```
# mount -t iso9660 /dev/cdrom /mnt/cdrom
```

Device Driver Module

The functional responsibility of the device driver is to read the external keypad and control the indicator LEDs.

1. Make a working directory on a local drive:

```
#mkdir /projects/DoorLock
```

2. Copy the source file from the companion CD:

```
#cp /mnt/cdrom/book/chapter9/DoorLock/pport.c  /projects/DoorLock/
```

3. Copy the Makefile file for the device driver module from the companion CD:

```
#cp /mnt/cdrom/book/chapter9/DoorLock/Makefile  /projects/DoorLock/
```

4. Copy the scripts to load and unload the device driver module:

```
#cp /mnt/cdrom/book/chapter9/DoorLock/pp_load  /projects/DoorLock/.
#cp /mnt/cdrom/book/chapter9/DoorLock/pp_unload  /projects/DoorLock/.
```

5. Create the loadable device driver module using the following command:

```
#cd /projects/DoorLock/dd
#make
```

6. Install the device driver using the script pp_load:

```
# ./pp_load
```

7. Verify that the device driver is in place using the following command:

```
# lsmod | grep pport
pport               1836    0 (unused)
```

Client/Server Module

The functional responsibility of the Client/Server module is:

- Communicate with the device driver and receive key entries from the user
- The coded keys are verified against the database
- If the key code is valid then open the door solenoid and turn on green LED
- Enter timestamp into the database

The test.c file and db.c file make up the Client/Server module.

ON THE CD

1. Copy the source file from the companion CD:

```
#cp /mnt/cdrom/book/chapter9/DoorLock/*.c  /projects/DoorLock/.
```

3. Copy the Makefile file for the module from the companion CD:

```
#cp /mnt/cdrom/book/chapter9/DoorLock/Makefile  /projects/DoorLock/.
```

4. Create the executable using the following command:

```
#cd /projects/DoorLock
#make
```

5. Execute the program with the following command:

```
# ./test
```

6. Verify that the module is in place using the following command:

```
# ps -ea | grep test
```

7. You should see the contents displayed as shown below

```
test        1836    0 (unused)
```

Web Server Module

The mini Web server developed in the *Sockets* section of Chapter 4 is integrated into the test.o file and the mini database manager described in Appendix B is being integrated into the db.c file. The test program launched in the previous section also launches the Web server that listens to the socket port number 9999. The next section describes how to view the contents of the database through the mini Web server.

Testing

The system test involves entering a key combination through the parallel port and viewing the entry log through the Web interface, as well as monitoring the parallel port status register pin-outs for logic levels indicating validation of the keys from the client/server module.

Testing the Key Combination

The switch combination for opening the lock is "1323" (see line 45 source listing cs.c).

1. Connect momentarily parallel port connector pin 9 with pin 15 (press sw1).
2. Connect momentarily parallel port connector pin 9 with pin 12 (press sw3).
3. Connect momentarily parallel port connector pin 9 with pin 11 (press sw2).
4. Connect momentarily parallel port connector pin 9 with pin 12 (press sw3).

The voltage level on pin 2 would be low, indicating that the solenoid is activated and door can be opened. For any other combination, the voltage level on pin 2 will remain high, while pin 3 will become low for three seconds.

Testing the Web Interface

1. Start Netscape Navigator and enter the URL:

```
http://localhost:9999
```

2. A Web page will be displayed as shown in Figure 9.7 showing the last 30 time-logs of the door-opening sequence.

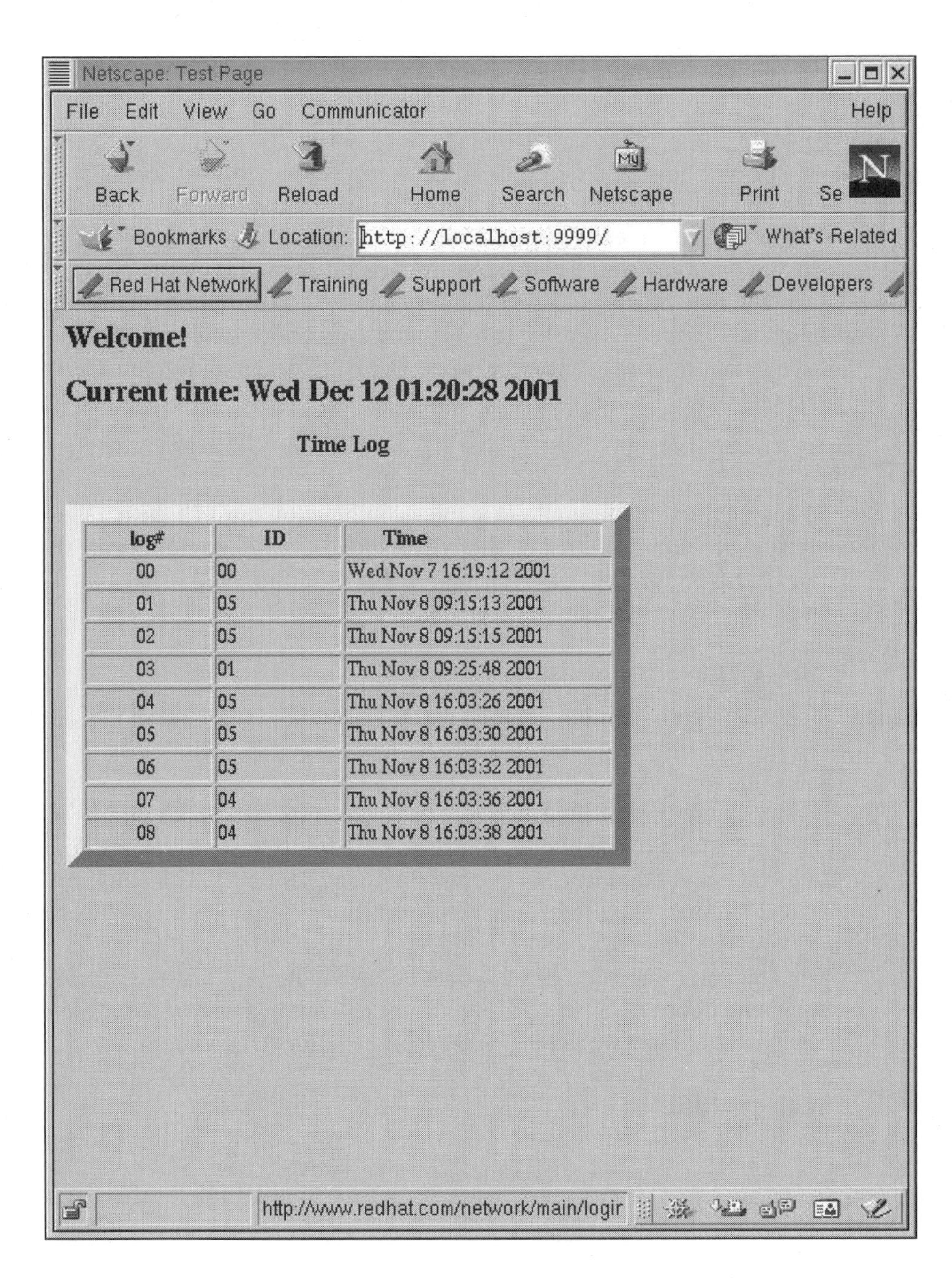

FIGURE 9.7 Entrance log display through Web browser.

PROJECT 2: AN ENHANCED ENTRANCE SECURITY SYSTEM WITH TIME LOG—DOORLOCK ENHANCED

A door-lock system with enhanced coded key and time log.

Product Specifications

- Low-cost coded-key door-lock system with enhanced time log facility
- Ideal for offices and manufacturing plants and hotel rooms with centralized control
- The time log database is available through the Web interface
- Web-based supervisory editing mode
- Built-in network facility for remote login
- Modular unit with built-in touch keypad and display

Introduction

This is an enhanced version of the DoorLock system; the basic features are same as in the previous project, with enhancement over the key combination, login facility of up to 1,000 entries, and supervisory control system via an intranet. It is a low-cost door-lock security system with a coded-key and time log facility, ideal for a hotel room locking system. The product will be developed using the Linux operating system, with the PC parallel port providing the necessary hardware circuitry.

The purpose of the implementation is to demonstrate the full-scale implementation of the Linux operating system, including the SQL, DNS, and HTTP servers, as a true Web application.

System Requirements

- Five-digit numeric keypad, simulated with the input pins of the status register of the PC parallel port
- Green LED signaling a valid key combination, controlled by the PC parallel data register bit 0
- Red LED signaling an invalid key combination, controlled by the PC parallel data register bit 1
- Lock solenoid latching signal, controlled by the PC parallel data register bit 2
- Capability of storing time log of the last 1,000 entries
- Capability of accepting multiple coded keys
- Supervisory access provided via a Web interface

The Hardware Functional Description

Please review the DoorLock Hardware Functional Description in the previous section.

The Software Functional Description

For a basic software functional description, please review the DoorLock Software Functional Description in the previous section. The enhanced version uses the transaction database server PostgreSQL and the Web interface through an HTTP/DNS server.

Functional Overview

The software modules for DoorLock are comprised of the system interface and the user interface.

The system interface modules:

- DeviceDriver
- ClientServer

The user interface modules:

- HTTP Server (Apache) configuration
- HTML pages and CGI scripts
  Index.html
  Supervisor.pl
- Database creation using the PostgreSQL server

The Device Driver Module

Please see the definition of the device driver module in the previous section, *DoorLock Functional Overview.*

The ClientServer Module

Please see the definition of the device driver module in previous section, *DoorLock Functional Overview.*

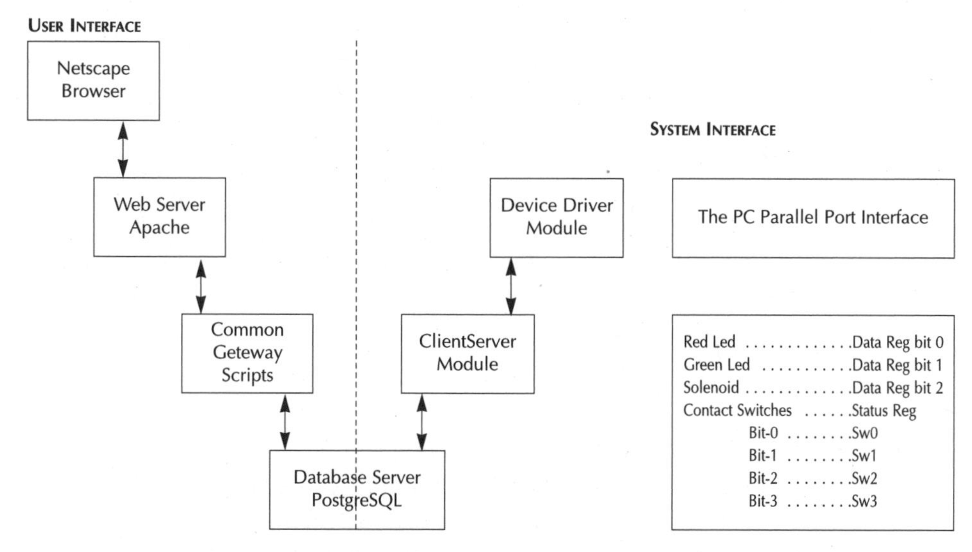

FIGURE 9.8 Diagram showing module interconnection for the enhanced DoorLock system.

The HTTP/DNS Server Configuration

Domain name, Index.html, and Perl script.

Please follow the instructions presented in Chapter 2 for setting up a Web site www.Timelog.com. The Web server module of the standard version of the Door-Lock is being replaced with the HTTP server module Apache. The supervisor can query the time log database via a Web browser and view the Web site at www.Timelog.com.

The index.html is essentially a Web form requesting the user to enter login name and a password for supervisory mode access. The Perl script log.pl is used to view the log-in information over the Web page.

Database Creation Using PostgreSQL

The PostgreSQL database server is used for the enhanced version of the DoorLock system. Please follow the directions specified in the procedure for creating a database in Chapter 2 using the following schema.

The Database Schema

The following tables are being set up for the enhanced version database setup:

Table Doorlock

Column 1	ID		(Char 6) —> Primary Key
Column 2	TimeIn	(TIMESTAMP)	

Implementation

ON THE CD

The module files are provided on the companion CD directory /book/chapter9/AdvancedDoorLock, and the contents should be copied to a local directory for compilation and testing purposes. Detailed instructions are also available in the ReadMe.txt file (CD directory path /book/chapter9/readme.txt) section "AdvancedDoorLock."

NOTE

You must be root user in order to perform the following operation.

Insert the companion CD into the CD-ROM drive and mount the CD-ROM using the following command:

```
# mount –t iso9660 /dev/cdrom /mnt/cdrom
```

Device Driver Module

The functional responsibility of the device driver is to read the external keypad and control the indicator LEDs.

1. Make a working directory on a local drive.

```
#mkdir /projects/AdvancedDoorLock/dd
```

ON THE CD

2. Copy the source file from the companion CD:

```
#cd /mnt/cdrom/book/chapter9/AdvancedDoorLock/dd
#cp pport.c  /projects/AdvancedDoorLock/dd/.
```

3. Copy the Makefile file for the device driver module from the companion CD:

```
#cp Makefile  /projects/AdvancedDoorLock/dd/.
```

4. Copy the scripts to load and unload the device driver module:

```
#cp pp_load  /projects/AdvancedDoorLock/dd/.
#cp pp_unload  /projects/AdvancedDoorLock/dd/.
```

5. Create the loadable device driver module using the following command:

```
#cd /projects/AdvancedDoorLock/dd
#make
```

6. Install the device driver using the script pp_load:

```
# ./pp_load
```

7. Verify that the device driver is in place using the following command:

```
# lsmod | grep pport
pport           1836       0 (unused)
```

Database Setup

1. Copy the database schema file from the companion CD:

```
#cd /mnt/cdrom/book/chapter9/AdvancedDoorLock/
#cp doorlock_schema /projects/AdvancedDoorLock/.
```

2. Switch to user postgres:

```
#su postgres
```

3. Create database doorlock:

```
$createdb doorlock
```

4. Start psql backend and populate the database using the doorlock_schema:

```
$psql doorlock
doolock-# psql \i doorloc
doolock-# \q
```

Client/Server Module

Similar to the DoorLock system of the previous section the functional responsibility of the Client/Server module is to:

- Communicate with the device driver and receive key entries from the user
- The coded keys are verified against the database
- Enter timestamp into the database

ON THE CD

1. Copy the source file from the companion CD:

```
#cd /mnt/cdrom/book/chapter9/AdvancedDoorLock
#cp *.c  /projects/AdvancedDoorLock/.
```

2. Copy the Makefile file for the device driver module from the companion CD:

```
#cp Makefile  /projects/AdvancedDoorLock/.
```

3. Create the client/server module:

```
#cd /projects/AdvancedDoorLock
#make
```

4. Execute the program with the following command:

```
# ./test
```

5. Verify that the test is being executed using the following command:

```
# ps -ea | grep test
```

6. You should see the contents displayed as shown here:

```
test       1836   0 (unused)
```

Web Hosting Setup

The AdvancedDoorLock system requires a virtual Web hosting setup with the domain name www.timelog.com.

Please follow the instructions presented in Chapter 2 for setting up the Web site www.Timelog.com.

ON THE CD

1. Copy the index.html file and Perl scripts from the companion CD:

```
#cp index.html /home/http/html/www/timelog/.
#cp *.pl /home/http/html/www/timelog/cgi-bin/.
```

Testing

The system test involves entering a key combination through the parallel port and viewing the entry log through the Web interface.

Testing the Key Combination

Follow the procedure described in the previous *DoorLock* section of *Testing the Key Combination.*

Testing the Web Interface

The AdvancedDoorLock Web interface provides more supervisory control than the previous implementation of the DoorLock.

Start Netscape Navigator and enter the URL:

```
http://www.timelog.com
```

The browser will display a page as shown in Figure 9.9. Enter "0000" for login and "password"' for password to gain access to supervisor mode. The browser will display a page showing the entry log.

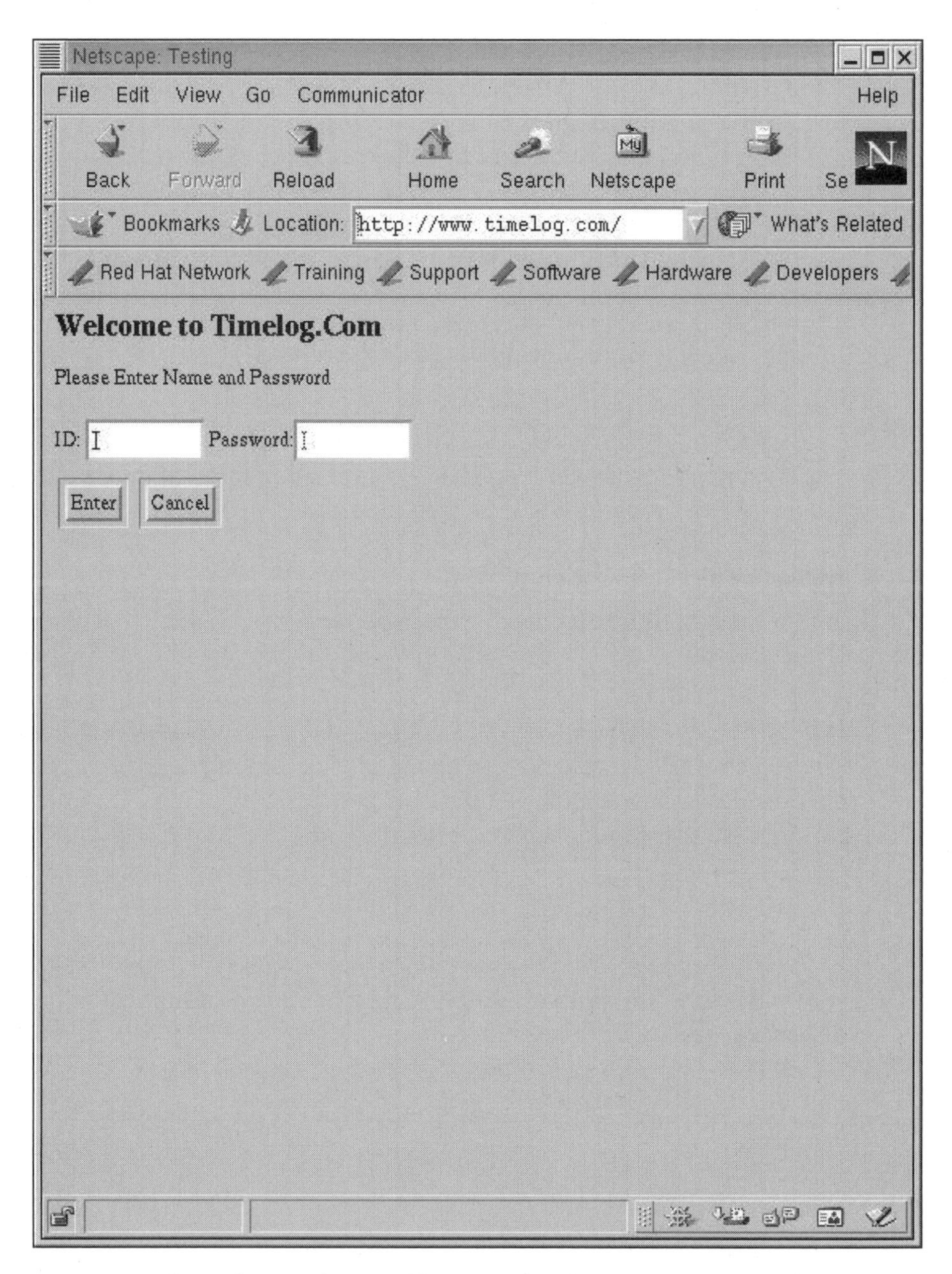

FIGURE 9.9 The Web page showing fill out form for login information.

Once the login is completed, you will be presented with the time log information page.

PROJECT 3: WHEN BEING ON TIME MATTERS—TIMEISMONEY

An integrated Web application for an employee time and attendance system.

Product Specifications

- Time and attendance application using the Web browser for time log entry
- Instantly retrieve record information for the last three years
- Supervisory mode viewing and editing of employee information
- Secured remote access of time log information
- Optional tax and payroll calculation
- Time log through the entrance security system
- Built-in network facility for distributed control
- Modular unit with built-in touch keypad and display

Introduction

Tracking employee time log and attendance is made easier with this integrated application that uses Linux as the operating system and a Web interface for user interaction. The modular system is an extension of the AdvancedDoorLock entrance security system described earlier. The enhancements are made so that staff members can log the time-in and time-out information through a Web interface.

In this project, we will enhance the concept of client/server communication developed in the previous section, and design a functional product that will use Web technology for the user interface, and an interprocess communication method such as message queues for the system interface. You will learn the basics of CGI scripting using Perl. The Perl modules Pg and CGI will be used to design the interface between a database server, such as PostgreSQL, and a Web browser, such as Netscape Navigator. With the help of these scripts, the user will be able to view the entrance time logs that are stored by the system modules, and be able to perform supervisory functions of editing the employee data and system environment.

The Hardware Functional Description

TimeIsMoney is basically a standalone module with a user interface through Web only. But it can also be used as an extension to the AdvancedDoorLock system, providing capability of entering access code through external keypad as well. Please review the *DoorLock Hardware Functional Description* in the previous section for adding the external key access.

The Software Functional Description

For a basic software functional description, please review the *DoorLock Software Functional Description* in the previous section. The database has been enhanced to incorporate more employee information for tracking the time log entry through supervisory control.

Functional Overview

The functionality of the enhanced DoorLock is incorporated with an additional user interface through Perl scripting, including supervisory mode access and employee login through the Web.

Enhanced User Interface Modules

- HTTP server (Apache) configuration
- HTML pages and CGI scripts
  Index.html
  Supervisor.pl
- Database creation using PostgreSQL server

The HTTP/DNS Server Configuration

The virtual Web hosting setup of the TimeIsMoney is identical to the Advanced-DoorLock system of the previous section. Please follow the instructions presented in Chapter 2 for setting up the Web site www.Timelog.com, if you have not already done so. The Web access provides supervisory as well user access for entering time-stamps and viewing time log history.

Database Creation Using PostgreSQL

The PostgreSQL database server is used for the TimeIsMoney system similar to the enhanced version of the DoorLock system. Please follow the directions specified in Chapter 2 for setting up the PostgreSQL Server, and create the database timelog.

Implementation

ON THE CD The module files are provided in the companion CD directory /book/chapter9/TimeIsMoney, and the contents should be copied to a local directory for compilation and testing purposes. Detailed instructions are also available in the ReadMe.txt file (CD directory path /book/chapter9/readme.txt) section "TimeIs-Money."

NOTE

You must be root user in order to perform the following operation.

Insert the companion data CD into the CD-ROM drive, and mount the CD-ROM using the following command:

```
# mount -t iso9660 /dev/cdrom /mnt/cdrom
```

Device Driver Module

The basic functionality of the TimeIsMoney does not require a device driver, since all the user entry is provided through the Web interface. However, if it is being used as an extension to the AdvancedDoorLock system, then you need to set up a device driver as described in the previous section.

Client/Server Module

The client/server module follows the pattern of the AdvancedDoorLock system described in the previous section.

Database Setup

The database timelog for TimeIsMoney has two tables, the employee table and the entrance_log table. The primary key ID of the employee table is a foreign key in the entrance_log table; thus, creating a relational database for the system. The PostgreSQL server must have been configured as described in Chapter 2 prior to this setup.

1. Make a working directory in the local drive:

```
#mkdir /projects/timelog/
```

2. Copy the database schema file from the companion CD:

```
#cd /mnt/cdrom/book/chapter9/timelog
#cp employee_schema /projects/timelog/.
```

2. Switch to user postgres:

```
#su postgres
```

3. Create database timelog:

```
$createdb doorlock
```

4. Start psql backend and populate the database using the doorlock_schema:

```
$psql doorlock
doolock-# psql \i employee_schema
doolock-# \q
```

User Interface Setup

ON THE CD

The Web provides the primary user interface for TimeIsMoney. It includes setting up the virtual Web hosting www.timelog.com similar to the AdvancedDoorLock system. Please follow the instructions presented in Chapter 2 for setting up the Web site www.Timelog.com, and copy the necessary scripts from the CD as described next.

1. Copy the index.html file and Perl scripts from the companion CD:

```
#cd /mnt/cdrom/book/chapter9/timelog
#cp index.html /home/http/html/www/timelog/.
#cp *.pl /home/http/html/www/timelog/cgi-bin/.
```

Testing

The system test involves entering a key combination through the parallel port if you have added the AdvancedDoorLock system features; otherwise, the test involves launching a Web browser and viewing the entry log through the Web interface.

Testing the Key Combination

The "TimeIsMoney" is simply an extension of the "AdvancedDoorLock" described earlier. In this implementation, the user can key in the ID through the simulated parallel port pins. Follow the procedure described in the previous *AdvancedDoorLock* section of *Testing the Key Combination.*

Testing the Web Interface

Launch your favorite Web browser (Netscape) and enter the URL http://www.timelog.com. The browser will display a page as shown in Figure 9.9. Enter "0000" for login and "password" for password to gain access to supervisor mode. The browser will display a new timestamp value as shown in Figure 9.10. You can add a new employee, view the employee timelog, and perform other supervisory functions such as changing name, password, and so forth.

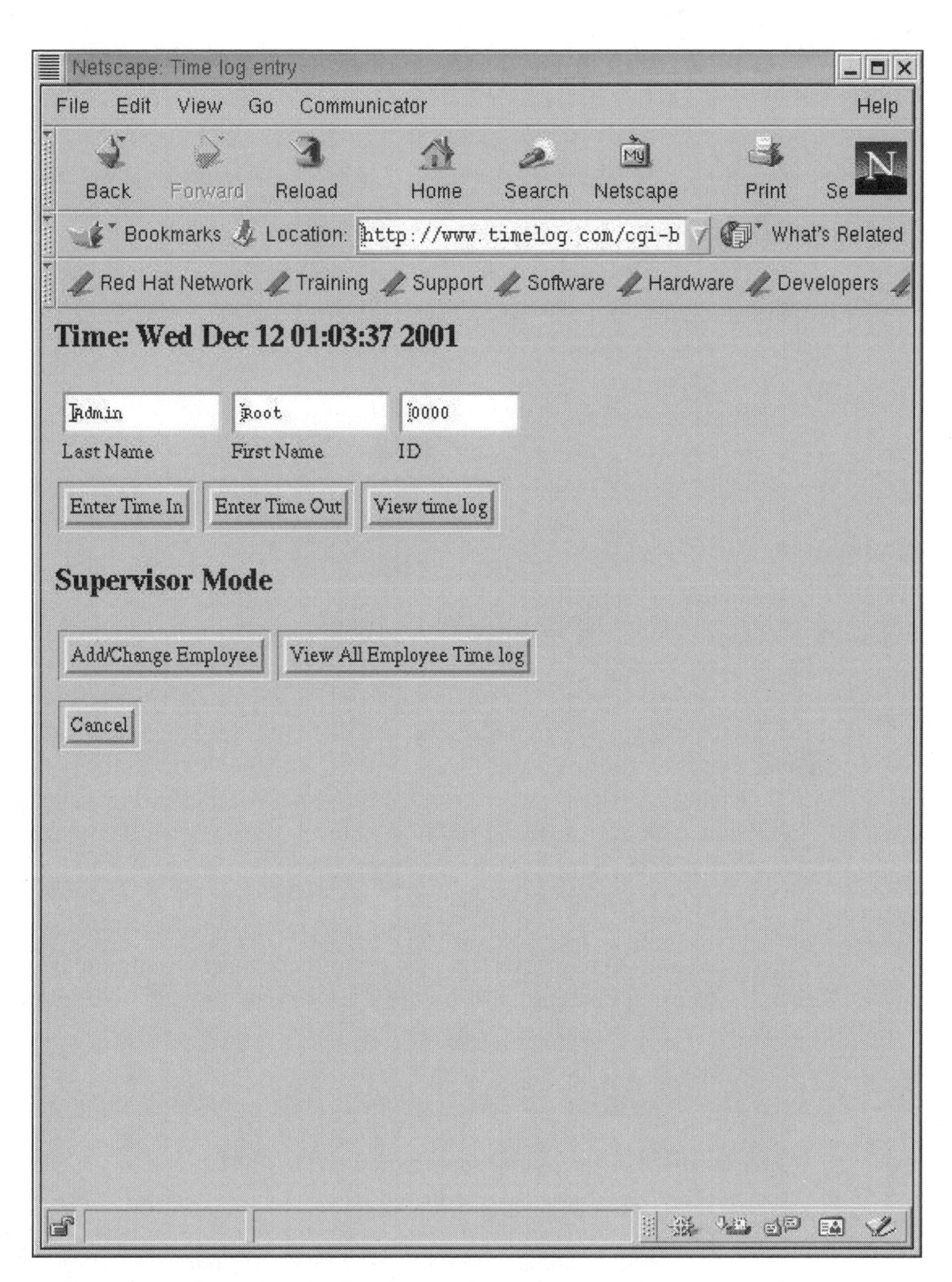

FIGURE 9.10 The Web page showing employee data entry form.

PROJECT 4: PLATFORM DRILLING RATE MONITORING SYSTEM WITH REMOTE LOG–DRILLINGRIG

Measuring oil platform drilling rate.

Product Specifications

- Low-cost data acquisition system with remote display
- Maintain history log of the entire duration
- Ideal for a network system with Web display and control
- Optional telemetry system available
- Optional programmable flow rate measurement
- Built-in network facility for remote access
- Modular unit with built-in touch keypad and display

Introduction

An important aspect of monitoring oil platform drilling activity is determining the depth of the prospective well and the rate of drilling. The most efficient way to acquire the desired signal is through monitoring the time movement of the cable attached to the drilling bits via a pulley mechanism. As the rig reaches a new depth, the cable advances to a certain distance, and by monitoring the time measurement angular movement of the pulley, we can determine the depth of the well and the rate of penetration. Measuring the angular movement of the pulley is accomplished by receiving an interrupt signal with every revolution of the pulley mechanism, similar to measuring a tachometer reading.

DrillingRig is a standalone data acquisition system designed specifically for monitoring the interrupt signals received from such a pulley movement. The data are processed and presented to the crew personnel for meaningful interpretation. The unit is also easily extensible as part of a network solution, since Ethernet and a modem port are provided with the system.

In this project, we will enhance the concept of client/server communication developed in Chapter 3, and design a product that will use Web technology for the user interface and interprocess communication method such as message queues for the system interface. You will learn the basics of CGI scripting using Perl. The Perl modules Pg and CGI will be used to design the interface between the database server (PostgreSQL) and the Web browser (Netscape Navigator). With the help of these scripts, the user will be able to view the digital signals received through the PC parallel port, and process the data as required for an oil platform drilling rate monitoring system.

The project uses the server push animation technique that allows the server to maintain an open connection with the client. The x-multi-replace MIME type automatically allows a periodic update of the Web page. The current value of the pulse counter is pushed to the client periodically, thus giving a near real-time refresh of the acquired data.

The Hardware Functional Description

The PC parallel port interrupt pin (DB-25 connector pin 10) is used as the channel for monitoring an external interrupt source, such as pulley movement of the drilling rig.

The Software Functional Description

The software functionality is divided into the system interface and a user interface. The user interface is basically a Web interface that provides continuous monitoring of the drilling activity via a Web page. The system interface, on the other hand, is the device driver module that receives external interrupts, and a client/server module that processes the data received from the driver.

Functional Overview

The functional block diagram and the interconnection of the software modules used in the system is similar to the DoorLock system as shown in the Figure 9.4.

The system interface modules:

- DeviceDriver
- ClientServer

The user interface modules:

- Web server setup
- DNS setup
- HTML/CGI scripts
  index.html
  drillingrig.pl

The Device Driver Module

The module communicates with the system hardware as a standard Linux device driver. A round-robin queue of MAX_PAGE entries is set up to store the timestamps of the interrupt pulse, received through the parallel port. The ioctl code

WAIT_FOR_INTERRUPT provides the pulse width between interrupts, while the client of the device driver is put on sleep until interrupt occurs. The following function calls must be supported:

int open(void)	Provides initialization of the hardware
int close(void)	Provides the closing of the file descriptor
int read(void)	Provides the keypad access
int write(void)	Provides the control access
int ioctl(void)	Provides the interrupt pulse width

Communication Protocol through the ioctl() Function

ioctl function call with WAIT_FOR_INTERRUPT parameter.

The ClientServer Module

This module is responsible for providing the desired system interface as a client/server model. The primary responsibility of this module is to receive the timestamp from the device driver module, and process the data as pulse width, pulse rate, and accumulated pulse rate, and store the values in the database.

The Web Server Setup

Please follow the instructions presented in Chapter 2 for setting up the Web site www.digital.com. The HTTP server Apache is being used as the primary Web server.

The DNS Server Setup

An entry is made to the DNS server for the domain www.drill.com.

Implementation

ON THE CD

The module files are provided on the companion CD directory /book/chapter9/drilling, and the contents should be copied to a local directory for compilation and testing purposes. Detailed instructions are also available in the ReadMe.txt file (CD directory path /book/chapter9/readme.txt) section "Drilling."

You must be root user in order to perform the following operation.

Insert the companion CD into the CD-ROM drive, and mount the driver using the following command:

```
# mount -t iso9660 /dev/cdrom /mnt/cdrom
```

Device Driver Module

The functional responsibility of the device driver is to provide interrupt pulse width.

1. Make a working directory on a local drive.

```
#mkdir /projects/drilling/dd
```

2. Copy the source file from the companion CD.

```
#cp /book/chapter9/drilling/dd/pport.c  /projects/drilling/dd/.
```

ON THE CD

3. Copy the Makefile file for the device driver module from the companion CD.

```
#cp /book/chapter9/drilling/dd/Makefile  /projects/drilling/dd/.
```

4. Copy the scripts to load and unload the device driver module.

```
#cp /book/chapter9/drilling/dd/pp_load  /projects/drilling/dd/.
#cp /book/chapter9/drilling/dd/pp_unload  /projects/drilling/dd/.
```

5. Create the loadable device driver module using the following command:

```
#cd /projects/drilling/dd/
#make
```

6. Install the device driver using the script pp_load.

```
# ./pp_load
```

7. Verify that the device driver is in place using the following command:

```
# lsmod | grep pport
```

8. You should see the contents displayed as shown in the following screen snapshot.

```
pport                1836    0 (unused)
```

Database Setup

1. Copy the database schema file from the companion CD:

```
#cd /mnt/cdrom/book/chapter9/drilling/
#cp drilling_schema /projects/drilling/.
```

2. Switch to user postgres:

```
#su postgres
```

3. Create database drilling:

```
$createdb drilling
```

4. Start psql backend and populate the database using the drilling_schema:

```
$psql drilling
doolock-# psql \i drilling_schema
doolock-# \q
```

Client/Server Module

ON THE CD

Please follow the script provided in the companion CD directory /book/chapter9/drilling/script.

1. Copy the source and Makefile files for the module from the companion CD.

```
#cp /book/chapter9/drilling/*  /projects/drilling/.
```

2. Create the executable using the following command:

```
#cd /projects/drilling
#make
```

3. Execute the program with the following command:

```
# ./test
```

4. Verify that the device driver is in place using the following command:

```
# ps –ea | grep test
```

5. You should see the contents displayed as shown here:

```
test        1836    0 (unused)
```

Web Hosting Setup

The DrillingRig system requires a virtual Web hosting setup with the domain name www.digital.com.

Please follow the instructions presented in Chapter 2 for setting up the Web site www.digital.com.

1. Copy the index.html file and perl scripts from the companion CD:

```
#cp index.html /home/http/html/www/digital/.
#cp *.pl /home/http/html/www/digital/cgi-bin/.
```

Testing

The Web interface is a push mechanism that periodically refreshes the depth value on the screen. The instantaneous depth value is obtained from the database. The database value is being updated by an interface through the device driver corresponding to an interrupt count received through the interrupt source of the parallel port.

Testing the Interrupts Received

1. Start your browser and enter www.digital.com.
2. Connect momentarily parallel port connector pin-9 with pin-10.
3. The browser will display a new timestamp value as shown in Figure 9.11.

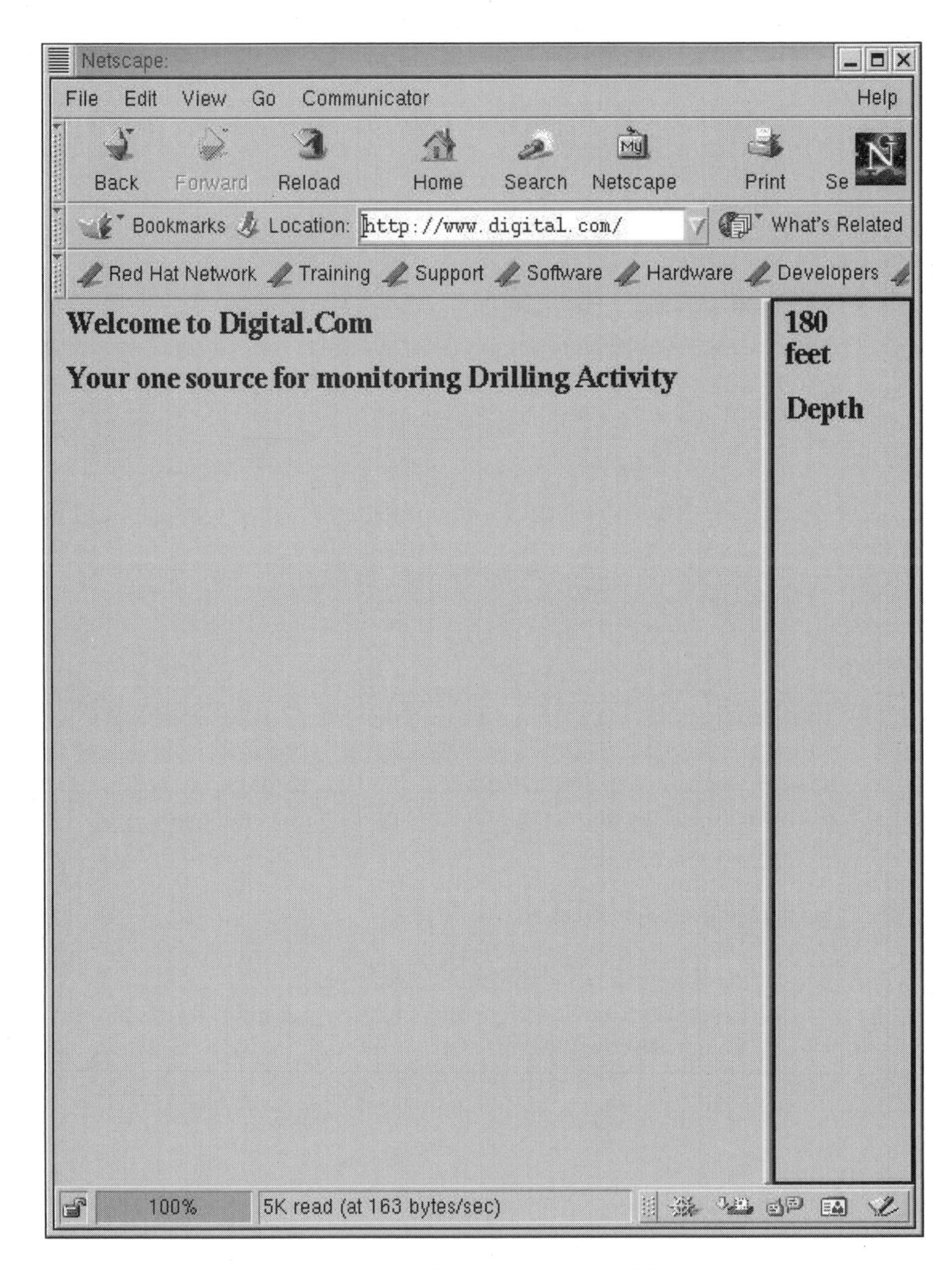

FIGURE 9.11 The Web page showing data from drilling activity.

PROJECT 5: LIQUID TANK LEVEL MONITORING SYSTEM–TANKVOLUME

A Web-based implementation of a liquid tank level monitoring system with optional input flow control through a pump shut-off system.

Product Specifications

- Low-cost data acquisition system with Web display
- Maintain history log
- Ideal for a network system with Web display and control
- Optional telemetry system available
- Optional programmable flow control through a pump shut-off system
- Built-in network facility for remote access

Introduction

An important aspect of an oil refinery system is maintaining the storage level of the tank farm for the production of the crude oil. An automatic pump shut-off system is desirable as part of a system solution. The liquid level in a tank can be monitored through installing contact switches at different height of the tank wall. The status of the switches indicates the level the liquid has reached in the tank. The system monitors the switches in real time, and displays the height on a Web page.

This project demonstrates the use of a server push animation technique that allows the server to maintain an open connection with the client. The x-multi-replace MIME type allows a periodic update of the Web page for a near real-time data display, similar to the one used in previous projects. For a graphic view of the tank level, the client Web page is refreshed periodically with new images.

The Hardware Functional Description

The PC parallel port status register pins are used as the contact switches for the height of the liquid level.

The Software Functional Description

The software functionality is divided into a system interface and a user interface. The user interface is basically a Web animation that provides continuous monitoring of the contact switches as the instantaneous value of the liquid level received from the database. The system interface, on the other hand, is the device driver module receiving external interrupts and a client/server module processing the data received from the driver and storing the value into the database.

Functional Overview

The functional block diagram and the interconnection of the software modules used in the system is similar to the DoorLock system as shown in Figure 9.4.

The system interface modules:

- DeviceDriver
- ClientServer

The user interface modules:

- Web server setup
- DNS setup
- HTML/CGI scripts
  index.html
  tank.pl
  welcome.html

The Device Driver Module

The module communicates with the hardware of the system as a standard Linux device driver. The height level is simulated through the status register of the parallel port. Each pin represents a contact closure at certain height.

The following function calls are supported:

int open(void)	Provides the initialization of the hardware
int close(void)	Provides the closing of the file descriptor
int read(void)	Provides the contact access
int write(int pattern)	Controls the logic level of the output pins

Communication Protocol through the read() Function

1. bit 0 = 0 for contact switch SW0.
2. bit 1 = 0 for contact switch SW1.
3. bit 2 = 0 for contact switch SW2.
4. bit 3 = 0 for contact switch SW3.
5. bit 4 = 0 for contact switch SW4.

The ClientServer Module

The module is responsible for providing the desired system interface as a client/server model. The primary responsibility of this module is to receive the contact switch information from the device driver module, process the data and store the values in the database.

The Web Server Setup

Please follow the instructions presented in Chapter 2 for setting up the Web site www.analog.com. The HTTP server Apache is being used as the primary Web server.

Implementation

The module files are provided on the companion CD directory /book/chapter9/TankVolume, and the contents should be copied to a local directory for compilation and testing purposes. Detailed instructions are also available in the ReadMe.txt file (CD directory path /book/chapter9/readme.txt) section "TankVolume."

You must be root user in order to perform the following operation.

Insert the companion data CD into the CD-ROM drive, and mount the driver g the following command:

```
# mount –t iso9660 /dev/cdrom /mnt/cdrom
```

Device Driver Module

1. Make a working directory on a local drive.

```
#mkdir /projects/TankVolume/dd
```

2. Copy the source file from the companion CD.

```
#cp /book/chapter9/TankVolume/dd/pport.c  /projects/TankVolume/dd/.
```

3. Copy the Makefile file for the device driver module from the companion CD.

```
#cp /book/chapter9/TankVolume/dd/Makefile  /projects/TankVolume/dd/.
```

4. Copy the scripts to load and unload the device driver module.

```
#cp /book/chapter9/TankVolume/dd/pp_load  /projects/TankVolume/dd/.
#cp /book/chapter9/TankVolume/dd/pp_unload  /projects/TankVolume/dd/.
```

5. Create the loadable device driver module using the following command:

```
#cd /projects/TankVolume/dd/
#make
```

6. Install the device driver using the script pp_load.

```
# ./pp_load
```

7. Verify that the device driver is in place using the following command:

```
# lsmod | grep pport
```

8. You should see the contents displayed as shown below.

```
pport              1836    0 (unused)
```

Database Setup

ON THE CD

1. Copy the database schema file from the companion CD:

```
#cd /mnt/cdrom/book/chapter9/tankvolume/
#cp drilling_schema /projects/tankvolume/.
```

2. Switch to user postgres:

```
#su postgres
```

3. Create database tankvolume:

```
$createdb tankvolume
```

4. Start psql backend and populate the database using the drilling_schema:

```
$psql drilling
doolock-# psql \i tankvolume_schema
doolock-# \q
```

Client/Server Module

ON THE CD

1. Copy the source file from the companion CD.

```
#cp /mnt/cdrom/book/chapter9/tankvolume/*  /projects/tankvolume/.
```

2. Create the executable using the following command:

```
#cd /projects/tankvolume
#make
```

3. Execute the program with the following command:

```
# ./test
```

4. Verify that the test is being executed using the following command:

```
# ps –ea | grep test
```

5. You should see the contents displayed as shown below.

```
test        1836   0 (unused)
```

Web Hosting Setup

The TankVolume system requires a virtual Web hosting setup with the domain name www.analog.com.

Please follow the instructions presented in Chapter 2 for setting up the Web site www.analog.com.

ON THE CD

1. Copy the index.html file and Perl scripts from the companion CD:

```
#cp index.html /home/http/html/www/analog/.
#cp *.pl /home/http/html/www/analog/cgi-bin/.
```

Testing

The Web interface is a push mechanism similar to the drilling monitoring system described in the previous section. The Web interface periodically refreshes the height of the liquid in the tank, using a live animation technique called a push mechanism. The instantaneous height value is obtained from the database. The database value is being updated by an interface through the device driver corresponding to the status port pins of the PC parallel port.

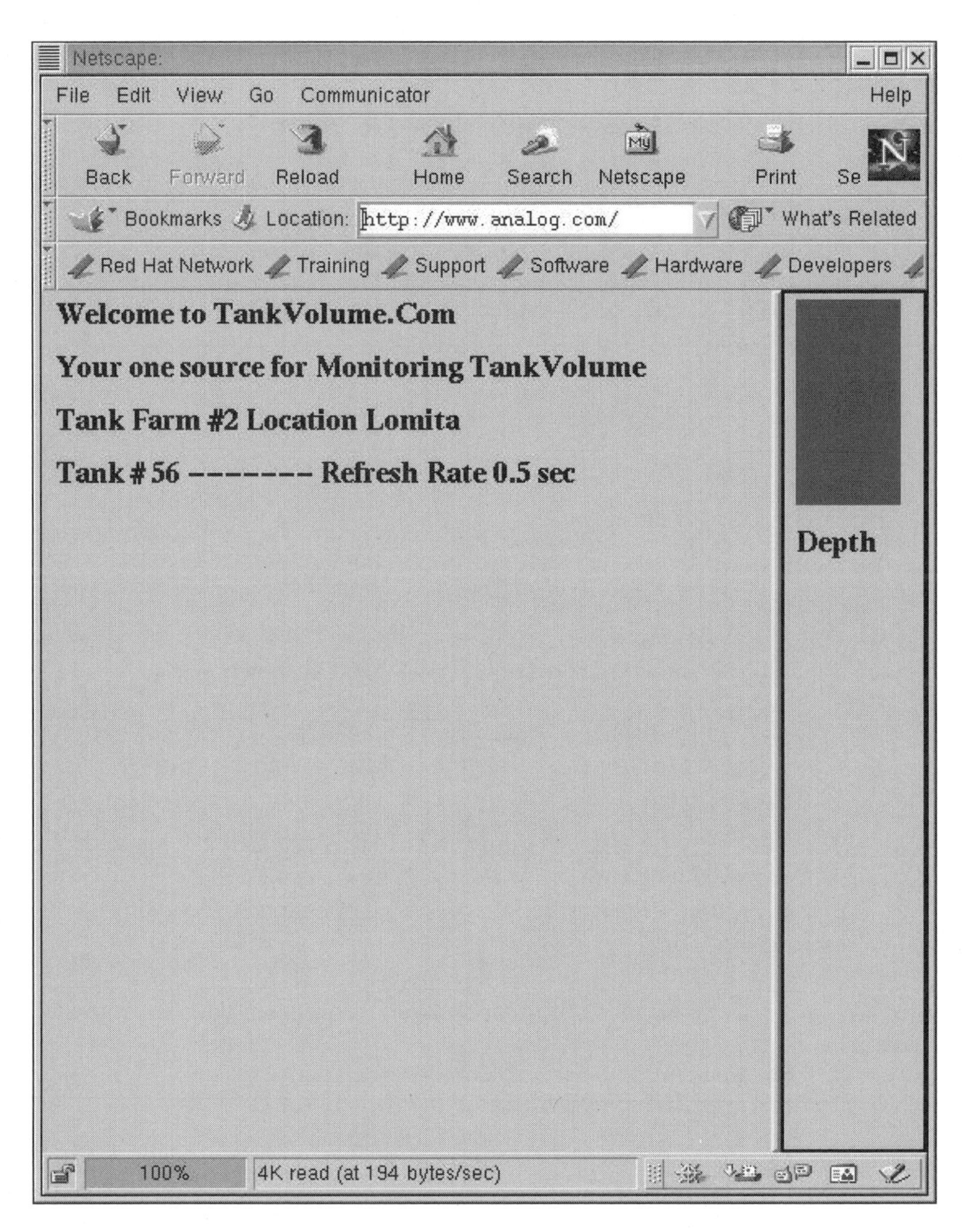

FIGURE 9.12 The Web page showing the liquid level height in the tank.

Testing the Interrupts Received

1. Start your browser, and enter www.analog.com.
2. Connect parallel port status register pins to ground.
3. The browser will periodically refresh the tank level on the screen as shown in Figure 9.12.

SUMMARY

In this chapter, we established the rules for designing embedded systems based on the concept of having a user interface, a system interface, and a database. Some practical design projects were implemented based on the design principle to prove that the concept is workable. Taking advantage of Web technology gave us a robust user interface with minimum effort. The device driver and client server mechanism for the system interface helped us achieve a modular design that is easily extensible. Having a database to link the user interface with the system interface minimized the interaction between the user and the system, and provided the persistence that is required for any practical embedded system design project.

Appendix

A C Programming Language Development Tools

The following sections are a summary of syntax rules.

VARIABLES AND IDENTIFIERS

Variables are objects that are being operated upon. A variable occupies a location in computer program memory and is identified by an *identifier*. An identifier is a character string of letters, digits, and underscores (_). The first character must be a letter. C is case sensitive; upper- and lowercase characters constitute different identifiers.

RESERVED WORDS

asm
auto
break
case
char
continue
default
do
double
enum

external
float
for
goto
if
int
long
register
return
short
sizeof
static
struct
switch
typedef
union
unsigned
void
while

VARIABLE STORAGE CLASS AND TYPE

A variable has two attributes associated with it: its *storage class*, that determines which section of the memory the identifier is located in, and the *type* that determines how to interpret the value.

Storage Class

Automatic: Local to each invocation of block; a new value is reassigned at each invocation.

Static: Local, but retains the initial value upon reentry.

External: Available to entire program, and retains the value for the life of the program.

Register: The compiler tries to use fast registers if possible. Should not be used as a parameter to pass to other functions.

Type and Constants

int: 32 bits of storage.

decimal integer constants: Sequence of digits 0..9, but does not begin with 0. For example:

123456

octal Integer constants: Sequence of digits 0..7, but begins with 0. For example:

0123

hexadecimal Integer constants: Sequence of digits 0..9 and a..f; begins with 0x or 0X. For example:

0xABCDEF

long: Constant is the largest integer 32-bits long. A decimal, octal, or hexadecimal constant followed by *l* or *L* is also considered long. For example:

0x1234l

short: Could be 16-bit or 32-bit long machine dependent. For example:

0x1234

char: 8-bit storage.

character constant: Is numerically equal to its value in the machine's character set, enclosed in single quotes such as 'd' is equivalent to 0x44. Some non-displayable characters can be defined as character constants if preceded with \, such as '\n' is equal to new line. For example:

'a' = 0x41

double: 64-bit storage.

float: 32-bit storage.

float and double constant: Consists of an integer portion and a fraction portion separated by a decimal point, preceded by an optional "-" sign. The number may be followed by an e or E followed by exponent value. For example:

-123.456E20

Enumeration constant sequence of names can be treated as numbers if defined as enum.

```
typedef enum printer_err
{
OUT_OF_PAPER,
INVALID_PAPER,
PAPER_JAM
}
```

OUT_OF_PAPER is treated as integer value 0.

INVALID_PAPER is treated as integer value 1.

PAPER_JAM is treated as integer value 2.

The enum does not have to be sequential; you can assign any constant to any member.

```
typedef enum printer_err
{
OUT_OF_PAPER = 3,
INVALID_PAPER = 7,
PAPER_JAM
}
```

OUT_OF_PAPER is treated as integer value 3.

INVALID_PAPER is treated as integer value 7.

PAPER_JAM is treated as integer value 8.

String literals: a sequence of characters surrounded by double quotes. For example:

"this is a string literal"

The compiler appends a NULL "\0" at the end of the string literal. The string literal has type array of characters and storage class *static*.

int and unsigned: The most significant bit of the *int* is treated as sign value, while the most significant bit of the *unsigned* is treated as part of the number. This affects the result of the compare operation, since unsigned values are always greater than or equal to zero, whereas the *int* can be less than, equal to, or greater than zero.

lvalues: The address of the object being referred; for example, if *int value* is a variable, then &*value* is the *lvalue*.

pointers: A variable exists somewhere in the program memory; the address of the location is a pointer. A pointer variable stores the address of another variable; thus, a pointer variable points to another variable. Addition and subtraction operations can be performed on pointer variables.

void: This is a catch-all phrase in C. A variable of type void has no type. No conversion is performed on void types.

EXPRESSIONS AND OPERATORS ON VARIABLES

Expressions are the arithmetic operations that are performed on variables and constants. The result generated by the operation can be used to assign new values to variables or tested for true or false condition in conditional statements such as *if* and *do while* statements.

Binary Arithmetic Operators

Operator	Description
+	Addition
-	Subtraction
*	Multiplication
/	Division
%	Modulus (produces remainder)

Unary Operators

Operator	Description
*	Pointer; the result is a pointer object, as in *char * test.*
&	lvalue, the result is the address of the object.
!	Not operator; the test compare with 0
~	1's complement, the result is 1's complement of the value.
++	Value, pre-increment by 1.
value++	Post-increment by 1.

Bitwise Operators

Operator	Description
&	Bitwise AND
|	Bitwise OR
^	Bitwise exclusive OR
<<	Left shift
>>	Right shift

Relational and Logical Operators

Operator	Description
>	Compare greater than, as in ... if (a > 2)
>=	Compare greater than or equal to, as in ... if (a >= 2)
<	Compare less than, as in ... if (a < 2)

<= Compare less than or equal to, as in … if (a <= 2)

= = Compare equal to, as in … if (a == 2)

!= Compare not equal to, as in … if (a != 2)

Precedence and Order of Evaluation

() [] -> Left to right

! ~ ++ — + - * & (type) sizeof Right to left

* / % Left to right

+ - Left to right

<< >> Left to right

< <= > >= Left to right

== != Left to right

& Left to right

^ Left to right

| Left to right

&& Left to right

|| Left to right

?: Right to left

= += -= *= /= %= &= ^= |= <<= >>= Right to left

, left to right

STATEMENTS

Operations that are executed in sequence.

Conditional Statements

if (expression) statement;

if (expression) statement *else* statements;

In the first instance, the expression is evaluated, and if the result is nonzero, the statement block is executed. In the second instance, the expression is evaluated, and if the result is nonzero, the statement block following the *if* is executed; otherwise, the *else* block is executed.

while Statement

while (expression) statement;

The expressions are evaluated and the statement block is executed, as long as the expression is true.

do while Statement

do (statement) while (expression);

The statement block is executed at least once, and then the expression is evaluated. The block is executed again until the expression is false.

for Statement

for (expression1; expression2; expression3) statements;

The expression1 is executed once, like an initialization. The statement block is executed as long as expression2 is true, and the expression3 is executed at the end of the statement block.

switch Statement

switch (expression)

case constant-expression : statements; break;

default : statements; break;

The switch statement causes control to be transferred to one of several statements depending on the expression matching one of the case constant-expressions. If no matching case is present, then statements followed by default are executed. The break terminates the statements of the switch block.

continue Statement

while (expression) { … continue; }

do { … continue; } while (expression) ;

for (exp1; exp2; exp3) { … continue; };

The continue statement causes control to be transferred back to beginning of the loop continuation portion.

break Statement

```
while (expression) { ... break; }
do { ... break; } while (expression) ;
for (exp1; exp2; exp3) { ... break; };
```

The break statement causes loop control to be terminated unconditionally.

return Statement

```
return;
return statement;
```

The return statement causes the control to be returned back to the caller. A value is returned back to the caller based on the return statement.

goto and label Statement

```
Testing: .... goto Testing
```

Any statement can be preceded by a label followed by a colon:. The goto statement transfers the control to the statement followed by the label. Label statements are only used by the goto statement.

Appendix

B Miscellaneous Linux Setup

This appendix provides information on the following Linux services:

- Point-to-Point Protocol (PPP)
- IP Masquerading
- Mail Client
- Printing
- Samba File and Print Server
- Dynamic Host Configuration Protocol (DHCP)
- vi Text Editor
- Basic Administrative Commands
- Database Manager (dbm)
- Network File System (NFS)

INTRODUCTION

Every computer environment is unique in some respect, and the various operating system services must be configured appropriately, based on the prevailing local requirements. For example, the ISP setup depends on whether you have a dial-up connection or a DSL, the security level might be shared or user level, the network might be required to forward IP requests, the printer connection might be local or remote, the mail is collected locally or left on the server, the Network File System (NFS) is allowed to share or be shared, and so forth. This appendix provides setup procedures and guidelines for various services that you may have installed as part of the system setup. It is not possible to cover all the different combinations available, so only a few basic options are discussed that might give you a head start in using the available services.

POINT-TO-POINT PROTOCOL

PPP allows you to connect to the outside world via a serial/modem link. The important point to remember is that some modems that use the *Host Processing technique* are not Linux compatible, such as WinModem or equivalent—you can tell by looking at the H sign on the modem box. They are inexpensive, but unfortunately some manufacturers have not provided drivers for Linux. External modems are guaranteed to work. The following setup is for Red Hat Linux version No, 7.1.

Components

Daemon '/usr/bin/pppd', 'inetd'

Use ifconfig to configure the interface.

Configuration

Prior to setup, you need to have the username and password from your ISP, and the DNS address and telephone number for connecting to the router.

Start the Control Panel.

Click "Big Foot" in the lower-left corner of the screen. Click System->Control Panel.

Modem Setup

Click Modem from the Control Panel, select "ttyS1." Click OK. (Assuming your modem is installed on COM port 2.) PC COM1 is ttyS0, and PC COM2 is ttyS1 on Linux.

PPP Setup

Click Network setup. Click "Interface'->'Add'->" Interface Type PPP. Type the telephone number of the ISP, and the login name.

Enter a password. Click on "Customize'->'Hardware Flow Control." Click "boot." Click "Communication." Edit the "Chat" Script. It should contain,

```
lock debug, modem noipdefault default route demand idle 300 holdoff 20
ipcp-accept-remote ipcp-accept-local
```

Verify

Verify the connection by clicking "Activate." You should hear the modem turn on and connect to the ISP. Catenate the log file "/var/log/messages" to make sure no PPP errors occurred during script processing.

IP MASQUERADING

IP masquerading allows you to share your one Internet connection with multiple machines. It is like sharing one IP address that was officially assigned by your ISP among several machines to communicate over the Internet. IP masquerading is essentially an IP forwarding by the machine that is directly connected to the Internet, with the IP address of the originating machine masqueraded with its own IP address. This allows other "internal" computers connected to the Linux box (via PPP, Ethernet, etc.) to reach the Internet, even though these internal machines do not have officially assigned IP addresses.

Components

IP masquerading is built into Linux. All the required features needed to support are compiled into it, such as:

- IPFWADM/IPCHAINS
- IP forwarding
- IP masquerading
- IP firewalling

Configuration and Setup

The file "'/etc/sysconfig/network" should contain:

```
FORWARD_IPV4=true
```

Add the following lines to the file "/etc/rc.d/rc.local":

```
echo "1" > /proc/sys/net/ipv4/ip_forward
echo "1" > /proc/sys/net/ipv4/ip_dynaddr
/sbin/depmod -a
/sbin/modprobe ip_masq_ftp
/sbin/modprobe ip_masq_raudio
/sbin/modprobe ip_masq_irc
/sbin/ipchains -M -S 7200 10 160
```

```
/sbin/ipchains -P forward DENY
/sbin/ipchains -A forward -s 192.168.0.0/24 -j MASQ
```

The preceding setup assumes you have selected 192.168.0.XX range of IP addresses for your local machines. You could also select any of the following addresses that the Internet Assigned Numbers Authority (IANA) has reserved for private networks:

10.0.0.0–10.255.255.255

172.16.0.0–172.31.255.255

192.168.0.0–192.168.255.255

MAIL CLIENT

Sending and receiving messages via the Internet.

Components

You need to contact your ISP for the machine names of the incoming and outgoing mail servers; most likely, they are the same.

The Linux daemon *sendmail* is the *MTA*, or Mail Transfer Agent, responsible for sending and forwarding messages. The configuration file for sendmail is "/etc/sendmail.cf".

Netscape Messenger, Pine, and Elm installed on Linux are the *MUA*, or Mail User Agent, that allows users to read and write e-mail.

SMTP (Simple Mail Transfer Protocol) is the sending protocol.

POP (Post Office Protocol) is the receiving protocol. The protocol allows you to retrieve the mail from the mail server and store it on your local machine. Once the mail is transferred, the server deletes the mail from its own archive.

MIME (Multipurpose Internet Mail Exchanger) encoded messages help retrieve images, etc.

Configuration and Setup

Outgoing Mail:

1. Click->Netscape->Edit->Preferences.
2. Click->Mail&NewsGroup->Identity.

3. In the form, specify your e-mail address and your e-mail return address.
4. Next, highlight the Mail Server option. Put the name of your SMTP server (the one provided by your ISP) in the "Outgoing Mail (SMTP)" field. Also fill in the appropriate username for that server.

Incoming Mail

1. Click->Netscape->Edit->Preferences.
2. Click->Mail&NewsGroup->Identity.
3. Click->Add. Enter the name of your incoming mail server in the server name. Select the protocol "POP" from the drop-down list box, and fill in the "User Name" field.

PRINTING

Set up local and network printing services.

Components

Daemon: /usr/bin/lpd

Print filters: magicfilter: download from ftp://metalab.unc.edu/pub/Linux/system/printing

Utility Files

/etc/printcap	Contains the printer specifications
/usr/bin/lpr	Submits a job to the printer
/usr/bin/lpq	Shows the contents of the spool directory
/usr/bin/lpc	Checks the status of the printer
/usr/bin/lprm	Removes a job from the printer

If your printer is a simple ASCII text printer, verify that "lpd" is present by printing the current directory to the printer with the following shell command:

```
#ls > /dev/lp0
```

Configuration and Setup

Suppose you have an HP LaserJet connected to the parallel port lp0, and you want all your print jobs to go to this printer. Append the following line to the file "/etc/printcap":

```
HP:\
      :lp=/dev/lp0:\
      :sd=/var/spool/lpd/:\
      :sh:\
      :mx#0:\
      :if=/root/magicfilter:
```

The options are colon separated: lp=/dev/lp0 is the line printer port, sd=/var/spool/lpd is the spool directory where jobs are spooled, sh is to suppress header, mx is the maximum job size (0 is unlimited), and if is the input filter. For more configuration options, enter "man printcap" at the shell command.

For a remote printer, append the following line to the file "/etc/printcap":

```
HP:\
      :lp=.:
      :sd=/var/spool/lpd/:
      :sh:
      :mx#0:
      :if=/root/magicfilter:
      :rm=remotehost:
      :rp=remoteprinter:
Reload 'lpd' by typing
killall -HUP lpd
```

Verify

You can verify your printer setup by pressing the printer button on your Netscape browser. Click on the File menu and select Print. Make sure the print command field is set to lpr –PHP.

SAMBA FILE AND PRINT SERVER

Making connections to Microsoft Windows computers.

Components

Daemon: inetd, /usr/bin/smbd, /usr/bin/nmbd

Configuration and Setup

1. Edit the file /etc/inetd.conf (you can use System->Application->gedit), and uncomment the line containing "swat" specifications.

```
swat
```

2. Click on the Command window (bottom row of your desktop), and restart inetd by typing in the following shell command:

```
#/etc/rc.d/init.d/inet restart
```

3. Start Netscape, and open the URL http://localhost:901/. When asked for User ID and Password, enter "root," "password." This will bring up the Samba configuration page.
4. Click on Globals.
5. Change the NetBIOS name to your machine name, such as ns1.
6. Click on Commit changes (upper-left corner).
7. Click on Status.
8. Click on Restart smbd, and Restart nmbd.

Verify

You can verify the presence of your machine on the network by opening up the Network Neighborhood on any Windows machine. You can also see Samba running. In a command window, enter the following shell command:

```
#smbclient – L ns1 –U root.
```

When prompted for Password, enter "password." It will show a list of shared volume and servers.

Sharing Your Printer with Windows Computers

If you have followed the printer setup defined in the previous section, you can share your Linux printer with other Windows machines on the network using the following Samba setup:

1. Add the following lines to the Samba configuration file "/usr/local/samba/lib/smb.conf":

```
1) [global]
2)    remote announce = 192.168.1.255
```

```
3)     interface = 192.168.1.1/255.255.255.0
4)     netbios name = ns
5)     workgroup = workgroup
6)     printing = bsd
7)     security = share
8)  [public]
9)     comment = public stuff
10)    path = /tmp
11)    public = yes
12)    writeable = yes
13) [printername]
14)    path = /var/spool/lpd
15)    printer name = HP
16)    writeable = yes
17)    public = yes
18)    printable = yes
19)   print command = lpr —PHP %s > /deb/lp1; rm %s
```

2. Restart Samba.

```
#killall —HUP nmbd; killall —HUP smbd
```

Make sure you can still browse files across the network. This time, you should see a printer icon with the assigned name from the "samba" configuration file.

Installing a Postscript Driver for a Win98 Client

Perform the following setup on your Windows machine:

1. From the desktop, Click Start->Setting->Printers.
2. In the Printer window, double-click Add Printers.
3. Click Next.
4. Click Network printers.
5. Click Next->Browse.
6. When you see your printer name, Click Ok->Next.
7. When you see the Add Printer Wizard dialog, select "HP" as the manufacturer and "LaserJet 4/4M PostScript" for the printer.
8. Restart Windows.

DYNAMIC HOST CONFIGURATION PROTOCOL (DHCP)

Allows client machines to get their IP address, gateway address, and DNS server address from the Linux server.

Components

Daemon: dhcpd

Configuration

/etc/dhcpd.conf

Setup

Assuming your domain IP address is 192.168.1.xx:

1. Edit the file /etc/dhcd.conf (you can use System->Application->gedit). It should contain:

```
default-lease-time 600:
max-lease-time 7200:
option subnet-mask 255.255.255.0:
option broadcast-address  192.168.1.255
option routers 192.168.1.1
option domain-name-server 192.168.1.1
option domain-name 'mydomain'
subnet 192.168.1.0 netmask 255.255.255.0 {
 range 192.168.1.10 192.168.1.100;
}
```

2. Edit the file /etc/rc.d/init.d/dhcpd (you can use System->Application->gedit). Add the following line in the Start section before dhcpd is called:

```
#route add –host 255.255.255.255 dev eth0
```

3. Touch (Create) the file /etc/dhcpd.leases by entering the following command at the shell prompt:

```
#touch /etc/dhcpd.leases
```

Verify

You can verify functionality by starting dhcpd manually by typing the following command at the shell prompt:

```
#dhcpd -d -f
```

Start any Windows machine and verify the IP address.

VI TEXT EDITOR

It is important to learn vi, as it is available in all mode of operations, whether you are in single-user mode with no X Windows, or in multiuser mode with a desktop environment such as K or GNOME. The vi is basically a full-screen editor with an extensive set of editing features, including search and replacement, block move and delete, cut and paste, multiple files, and so forth.

vi has basically two modes of operations, input mode and command mode. In input mode, the characters you type in are treated as text. In command mode, the characters such as a, A, i, I, x, and so forth are treated as commands to navigate around the screen. You can exit input mode by simply pressing the [ESC] key, and get in back in to input mode by pressing any of the special keys a, A, i, I, and so forth. Pressing ":" will drop the cursor to the bottom of the screen, and vi will wait for your command such as block move, search, write to file, quit, and so forth.

Starting vi

To edit a file named test.c, start a shell and enter the following command at the prompt:

```
$ vi file.c
```

You are in command mode.

Insert or Delete Text

i For insert mode, the character you type will show up as text.

x For delete character.

3x For delete three consecutive characters.

a For append character.

[ESC] To get into command mode.

Undoing Changes

u	To undo the most recent change.
U	To undo all the changes made to the line.
P	To put the contents of the recent buffer back to where the cursor is.
:e!	Throw away all changes and edit again.

Quitting vi

Press [ESC] **: wq** to write the file to disk and quit.

Press [ESC] **: q!** to quit without writing the file to disk.

Navigating around the Window

l or **spacebar** or **right arrow**	To move right one character.
h or **CTRL-H** or **left arrow**	To move left one character.
j or **CTRL-J** or **CTRL-N** or **down arrow**	To move down one line.
k or **CTRL-P** or **up arrow**	To move up one line.
0	To move to the top of the file.
$	To move to the end of the current line.
+ or **RETURN**	To move to the beginning of the next line.
-	To move to the beginning of the previous line.

Screen Movements

CTRL-F	Move forward one full screen.
CTRL-D	Move forward one-half screen.
CTRL-B	Move back one full screen.
CTRL-U	Move back one-half screen.
G	Move to the end of the file.
35G	Move to the 35th line.

Replacing Text

rb	Replace the current character with b.
3rb	Replace three characters with b.

Searching

/look	Search forward for the first occurrence of look.
//	Repeat search.
n	Repeat search.
N	Reverse search direction.
?look	Search backward for the first occurrence of look.

Copying and Moving Text

2yw	Yank the next two words and save them in a buffer.
yy	Yank the current line and save it in a buffer.
Y	Yank the current line and save it in a buffer.
3Y	Yank the following three lines and save them in a buffer.
p	Put the yanked text at cursor position.
P	Put the yanked text before the cursor position.

Working Buffers

There are 26 working buffers, a through z. To use them, you begin the command with double quotes ".

"t3Y	Yank the following three lines and save them in t buffer.
"T3Y	Yank the following three lines and append them to t buffer.
"tP	Put the contents of the t buffer back into text.

Editing Multiple Files

```
vi filea.c fileb.c filec.c
```

The vi starts with filea.c.

"a5Y	Yank five lines from the text and save them in the a buffer.
:n	Switch to next file fileb.c.
"aP	Paste the buffer from filea.c into fileb.c.

Inserting Output from the Shell Command

:r !date	Read the output of date command, and place the buffer at the current line.
:r !ls	Read the output of ls command, and place the buffer at the current line.

BASIC ADMINISTRATIVE COMMANDS

Linux administrative commands involve adding and removing user and group accounts and passwords, and system maintenance commands such as the disk, directory usage, and so forth.

Adding a User

Enter the following line to add a user name xtapo. The –m option will create a home directory in /home/xtapo.

```
#useradd –m xtapo
```

useradd Options

-u uid	Sets the user id.
-o	Assigns a uid that is not unique.
-g group	Assigns the user to the group.
-d dir	Sets the home directory.
-e expire	sets the expiration date; the default is 0 (no expiration).

Deleting a User

Enter the following line to delete the user xtapo; -r is to remove the home directory:

```
#userdel –r xtapo
```

Adding a Group

Enter the following line to add a group named developers:

```
#goupadd developers
```

Deleting a Group

Enter the following line to delete the group developers:

```
#goupdel developers
```

User Password

Enter the following line to modify the password for the user xtapo:

```
#passwd xtapo
```

Display System Name

Enter the following command to display all system name information:

```
#uname –a
```

Display Disk Space

Enter the following command to see the amount of disk space available in each file system:

```
#df –t
```

Display Disk Usage

Enter the following command to see the directory space /var/named:

```
#du /var/named
```

Display Processes Running on the System

Enter the following command to see all processes running on the system:

```
#ps -ef
```

DATABASE MANAGER

Some embedded system designs might have space constraints, so implementing a full-scale database server such as PostgreSQL may not be feasible. Linux offers a scaled-down version of database manager called dbm. The dbm is essentially a library of routines that create and maintain hash tables, with each entry as a key and value pair. The key acts as an index into the data stored as if it was a label for identification. The key must be unique, since no duplicates are allowed.

The concept of dbm is very simple: you provide the key and the data; the dbm provides only basic routines for storing, retrieval, and deletion. There is also a way of traversing the database, but in a non-sorted way. There is no restriction on types of data as well as the structure of the key that can be stored by the dbm. The only information needed is the pointer to the key and the size of the key, and a pointer to the data and the size of the data, as defined by the typedef *datum* in the header file ndbm.h.

```
typedef struct {
    char *dptr;
    int   dsize;
    } datum;
```

The header file also defines a DBM structure as follows:

```
typedef struct {int dummy[10];} DBM;
```

The DBM is similar to a file structure, as the same basic four functions of open, close, read, and write are provided, plus an extra delete function to complement the write is applied on the database as well. The corresponding read, write, open, and close functions in the dbm implementation are dbm_open, dbm_close, bm_fetch, and dbm_store. Deleting a record from the database is accomplished through dbm_delete. The following are the basic functions defined in the header file ndm.h for dbm access:

dbm_open

```
DBM *dbm_open (char * name, int flags, int mode);
```

The database file must be created before it can be used. The mode parameter:

```
O_RDWR | O_CREATE
```

will create the database. Any filename is valid as long as the extensions .dir and .pag are not used. The dbm creates two working files with the extensions .dir and .pag for its own internal use. The dbm_openreturns a pointer to DBM type if it is successful; otherwise, a NULL is returned. The returned pointer should be used for subsequent access.

dbm_store

```
int dbm_store (DBM * file, datum key, datum content, int flags);
```

The function writes the key and the contents to the database. The DBM * file is the returned value from the dbm_open function. The flag can either be dbm_insert or dbm_replace. The dbm_insert fails if the data already exist and returns 1; otherwise, a successful data insertion returns 0. For dbm_replace, a successful replacement returns 0; all other errors return a –1.

dbm_fetch

```
datum dbm_fetch (DBM * file, datum key);
```

The function searches for the key and retrieves the value. The returned datum.dptr element will be NULL if the search fails; otherwise, the datum.dptr has the pointer to the value, and the datum.dsize has the size of the value.

The actual data must be copied to the local memory before any other call to the dbm is being made.

dbm_delete

```
datum dbm_delete (DBM * file, datum key);
```

The function deletes the key and the value pair from the database. The 0 is returned if there was a successful delete.

dbm_close

```
datum dbm_close (DBM * file);
```

The function closes the open database (DBM * file) that was opened with dbm_open.

Navigating around the Database

The dbm also provides support functions that help in moving around the database, so that each record in the database can be accounted for.

dbm_firstkey

```
datum dbm_firstkey (DBM * file);
```

The routine positions the database cursor at the beginning of the file. From a programming point of view, it is only a reference point, as records are not stored in any sorted order. If the returned value datum's member datum.dptr is NULL, then the database is empty.

dbm_nextkey

```
datum dbm_nextkey (DBM * file);
```

The routine moves the database cursor to the next position in the database file. If the returned value datum's member datum.dptr is NULL, then the cursor has reached the last record in the database. The following lines compute the total number of records in the database:

```
DMB *dp_ptr
datum key, value;
int total_record = 0;
for (key = dmb_firstkey(db_ptr); key.dptr; key = dbm_nextkey(db_ptr))
total_record++;
```

dbm_error

```
datum dbm_error (DBM * file);
```

The routine simply tests for errors in the database. Return 0 indicates no error occurred.

dbm_clearerr

```
datum dbm_clearerr (DBM * file);
```

The routine clears any error that occurred during prior operations on the database.

NETWORK FILE SYSTEM

NFS is available on several different platform including, AIX, Sun Solaris, and Windows, but it is only used in file and directory sharing. The device sharing is not supported on NFS. With NFS, the remote systems can mount the resources that are being shared among systems. Although Samba may be a better choice for sharing with Windows systems, knowing Microsoft you cannot rely upon the technology to be compatible across platforms. In fact SMB security has been changed in Windows NT 4.0 service pack 3 that makes Samba inoperable. In any event NFS is a better choice for sharing files among UNIX systems. The Linux homepage for nfs is at http://nfs.sourceforge.net.

NFS related man pages are nfs(5), exports(5), mount(8), fstab(5), nfsd(8), lockd(8), statd(8), rquotad(8), and mountd(8).

NFS Server Installation

The following daemons make up the NFS services:

rpc.nfsd	The main service daemon.
rpc.lockd and **rpc.statd**	Handle file locking.
rpc.mountd	Handles the initial mount requests.
rpc.rquotad	Handles user file quotas on exported volumes.

Most recent Linux distributions have startup scripts for these daemons. The daemons are part of the nfs-utils package, and can be either in the /sbin directory or the /usr/sbin directory. NFS depends on the portmapper daemon, either called portmap or rpc.portmap. You can see the portmapper running with the following command:

```
ps aux | grep portmap
```

If you have already selected NFS at installation time, you can verify the daemon with the following query to portmapper:

```
rpcinfo -p
```

If the services are running, you will see the following message:

```
program vers proto   port
100000    2   tcp    111   portmapper
100000    2   udp    111   portmapper
100024    1   udp    944   status
100024    1   tcp    946   status
100005    1   udp    759   mountd
100005    1   tcp    761   mountd
100005    2   udp    764   mountd
100005    2   tcp    766   mountd
100005    3   udp    769   mountd
100005    3   tcp    771   mountd
```

The Linux systems use UDP by default unless TCP is explicitly requested; however, other operating systems such as Solaris default to TCP. If the NFS daemon is not present in the /sbin or /usr/sbin, you can install it from the distribution CD. For the Red Hat version, find out which NFS module is provided in the distribution with the following shell command:

```
rpm –qa | grep nfs
```

Install the version with the following command:

```
rpm –i nfs-version
```

Created a startup script using the following chkconfig command:

```
chkconfig - - add nfs
```

Server Configuration File Setup

The following three configuration files provide the configuration information for the NFS server **/etc/exports**

This file contains a list of entries, indicating which volume is shared. See the man pages (man exports) for a complete description of all the setup options for the file.

The format for entry is as follows:

```
directory IPAddress1(option11,option12) IPAddress2(option21,option22)
```

The *directory* indicates the shared directory, and the IP addresses are for the client machines that will have access to the directory. The machines can be listed by their IP address or their DNS address. The following options are recognized:

+ **ro** The directory is shared read only; the client machine will not be able to write to it. This is the default.

+ **rw** The client machine will have read and write access to the directory.

+ **no_root_squash** This is the most important option, as it affects the root. By default, root access is treated as "nobody"—literally, the user "nobody," but with no_root_squash, the root permissions are granted.

+ **no_subtree_check** The subtree checking verifies that a file that is requested from the client is in the appropriate part of the volume.

+ **sync** The option forces the file system to sync to disk every time NFS completes a write operation.

A typical setup for /etc/exports might look like this:

```
/usr/local   192.168.1.20/255.255.255.0 (ro) ns1.aklinux.com(ro)
```

/etc/hosts.allow and /etc/hosts.deny

These two files specify which computers on the network can use services on your machine. The server first checks hosts.allow to see if the machine matches a description listed there. If it does, the machine is allowed access. If the machine does not match an entry in hosts.allow, the server then checks the hosts.deny to see if the client matches a listing there. If it does, the machine is denied access. If the client matches no listings in either file, it is allowed access.

The following is the format for each entry in the file:

```
service: host [or network/netmask] , host [or network/netmask]
```

It is generally a good idea to explicitly deny access to all hosts. The following entry in the /etc/hosts.deny file accomplishes the task:

```
portmap:ALL
lockd:ALL
mountd:ALL
rquotad:ALL
statd:ALL
```

We then add an entry to hosts.allow to give client access. Entries in hosts.allow have the same format as that of host.deny files.

```
service: host [or network/netmask] , host [or network/netmask]
```

The following entry to /etc/hosts.allow gives access to two machines, 192.168.0.1 and 192.168.0.2:

```
portmap: 192.168.0.1 , 192.168.0.2
lockd: 192.168.0.1 , 192.168.0.2
rquotad: 192.168.0.1 , 192.168.0.2
mountd: 192.168.0.1 , 192.168.0.2
statd: 192.168.0.1 , 192.168.0.2
```

NFS CLIENT SETUP

NFS clients can mount directories that are exported by an NFS server. You can use the utility "showmount" with the NFS server hostname or IP address to see what directories are being exported by the server. For example:

```
#showmount 192.168.0.40
```

Mounting Remote Directories

You need to have a kernel with built-in NFS support. You can verify this by looking at a listing of the /proc files system. You should see a line with nfs; otherwise, you need to rebuild the kernel with NFS support or download one with the support built in. To begin using a machine as an NFS client, you need the portmapper running on that machine. To use NFS file locking, you will also need rpc.statd and rpc.lockd running on both the client and the server.

The following mount command will mount the home directory of the machine 192.168.0.1 to /mnt/home of the local machine:

```
# mount 192.168.0.1:/home /mnt/home
```

Use umount to remove the mount

```
# umount /mnt/home
```

SUMMARY

In this appendix, we have presented guidelines on how to set up various Linux services, including the Point-to-Point Protocol (PPP), IP Masquerading, Mail Client, Printing, Samba File and Print Server, Dynamic Host Configuration Protocol (DHCP), and Network File System (NFS). We also discussed how to use different utilities, including the vi Text Editor, Basic Administrative Commands, and the Database Manager (DBM).

Appendix

C HTML Tags

In this appendix

- HTML Tag Attributes
- HTML Tag Definitions

INTRODUCTION

To complete the discussion of the user interface design for our embedded systems, we need to understand HyperText Markup Language (HTML) pages. A text defined in HTML is what you get when you open a browser and request a Web site. The HTML tags are special bracketed string tokens that are interpreted only by browsers such as Netscape and Internet Explorer. The browser creates special effects on the Web page based on the HTML tags, and controls the formatting and layout of HTML pages; the user provides the tags. The tags are input from the user, providing a set of functionality that needs to be applied, and the browser interprets the tag and sets the layout of the page before it is displayed.

You are not expected to apply these tags manually, as there are several elegant HTML editors available such as FrontPage and Dreamweaver. However, as an embedded system designer, you might have to create a Web page dynamically, one tag at a time, as if it was created by an HTML editor. The only tool you have is the executable created by a C compiler using printf routines, outputting the desired string tokens and creating the effect on the user screen as if the page was created by an HTML editor. In this appendix, we present a comprehensive description of HTML tags with examples to show the string of characters required to generate the desired effect. The World Wide Web consortium (W3C) is the custodian of the Internet standard and specifies the HTML and XML basic protocols, including the behavior of the tags.

HTML TAG ATTRIBUTES

HTML tags are essentially functions that decompose an HTML page and affect how a page is displayed or interpreted; for example, the <BGSOUND> tag is a function that activates a sound file. The function takes attributes such as a LOOP count that defines how many times the file is to be played. The attributes to HTML tag functions specify the characteristics of the function. If <P> is a tag that formats a paragraph, then the ALIGN attribute, whose values LEFT, RIGHT, CENTER, or JUSTIFY, affects the appearance of the paragraph. The following are the common attributes for most of the HTML tag functions,

ALIGN={LEFT | RIGHT | CENTER | JUSTIFY } Defines the placement of the text on the screen.

CLASS="..." Indicates the style class to be applied to the tag elements, such as <P>, <H1>, <UL>, and so forth. See the definition of *style* in the next section.

COLOR="rrggbb" The attribute value rrggbb is a six-digit hexadecimal number; the first two digits specify the red value, the middle two the green value, and the last two the blue value. Some sample color values:

red FF0000, green 00FF00, blue 0000FF, black 0, white FFFFFF, gray 888888 yellow FFFF00, cyan 00FFFF

Microsoft Explorer supports the use of color names in the BGCOLOR, BORDERCOLOR, BORDERCOLORLIGHT, and BORDERCOLORDARK attributes. Valid colors are

Aqua, B lack, Blue, Fuchsia, Gray, Green, Lime, Maroon, Navy, Olive, Purple, Red, Silver, Teal, Yellow, and White.

SHAPE Can be one of RECT, CIRCLE, POLY, or DEFAULT. CO-ORDS gives the coordinates in pixels, measured from the upper-left corner of the image being referenced for the shape. For RECT, these are left, top, right, and bottom; for CIRCLE, they are Xcenter, Ycenter, radius; and for POLY, they are x1, y1, x2, y2, ... xn, yn.

STYLE="..." Specifies the style sheet commands that apply to the contents of the tag elements. Multiple style definitions may be separated by semicolon; for example:

<P STYLE="background: red; color: white">

You can choose from one of the following categories of style properties:

- Font
- Text
- Box
- Color and background
- Classification
- Printed style sheet
- Positioning

Table C.1 is an alphabetical list of HTML tokens and the functions they represent.

TABLE C.1 HTML Tag List

Tags	*Description*	*Tags*	*Description*
<!– –>	Comment	<IMG>	Inline Image
<!DOCTYPE>	Escape Sequences	<INPUT>	Form Input
<A>	Anchor	<INS>	Inserted Text
<ABBR>	Abbreviation	<KBD>	Keyboard
<ACRONYM>	Acronym	<LANG>	Language
<ADDRESS>	Address	<LH>	List Heading
<APPLET>	Java Applet	<LI>	List Item
<AREA>	Anchor	<LINK>	Link
<AU>	Author	<LISTING>	Listing
<AUTHOR>	Author	<MAP>	Map
<B>	Bold	<MARQUEE>	Marquee
<BANNER>	Banner	<MATH>	Math
<BASE>	Base	<MENU>	Menu List
<BASEFONT>	Base Font	<META>	Meta
<BGSOUND>	Background Sound	<MULTICOL>	Multi Column Text
<BIG>	Big Text	<NOBR>	No Break
<BLINK>	Blink	<NOFRAMES>	No Frames
<BLOCKQUOTE>	Block Quote	<NOTE>	Note
<BQ>	Block Quote	<OL>	Ordered List

Tags	Description	Tags	Description
<BODY>	Body	<OVERLAY>	Overlay
 	Line Break	<P>	Paragraph
<CAPTION>	Caption	<PARAM>	Parameters
<CENTER>	Center	<PERSON>	Person
<CITE>	Citation	<PLAINTEXT>	Plain Text
<CODE>	Code	<PRE>	Reformatted Text
<COL>	Table Column	<Q>	Quote
<COLGROUP>	Table Column Group	<RANGE>	Range
<CREDIT>	Credit	<SAMP>	Sample
<DEL>	Deleted Text	<SCRIPT>	Script
<DFN>	Definition	<SELECT>	Form Select
<DIR>	Directory List	<SMALL>	Small Text
<DIV>	Division	<SPACER>	White Space
<DL>	Definition List	<SPOT>	Spot
<DT>	Definition Term	<STRIKE>	Strikethrough
<DD>	Definition Definition	<STRONG>	Strong
<EM>	Emphasized	<SUB>	Subscript
<EMBED>	Embed	<SUP>	Superscript
<FIG>	Figure	<TAB>	Horizontal Tab
<FN>	Footnote	<TABLE>	Table
<FONT>	Font	<TBODY>	Table Body
<FORM>	Form	<TD>	Table Data
<FRAME>	Frame	<TEXTAREA>	Form Text Area
<FRAMESET>	Frame Set	<TEXTFLOW>	Java Applet Textflow
<H1>	Heading 1	<TFOOT>	Table Footer
<H2>	Heading 2	<TH>	Table Header
<H3>	Heading 3	<THEAD>	Table Head
<H4>	Heading 4	<TITLE>	Title
<H5>	Heading 5	<TR>	Table Row
<H6>	Heading 6	<TT>	Teletype
<HEAD>	Head	<U>	Underlined

Tags	Description	Tags	Description
<HR>	Horizontal Rule	<UL>	Unordered List
<HTML>	HTML	<VAR>	Variable
<ITALIC>	Italic	<WBR>	Word Break
<IFRAME>	Frame - Floating	<XMP>	Example
<ISINDEX>	Is Index		

HTML TAG DEFINITIONS

The following common tags are supported by both Netscape and Internet Explorer:

<!–>

Comment Includes the actual commented text. Any instance of —> ends the comment. Whitespace can be included between the — and the >, but not between the <! and the first —.

```
<!– comment text –>
```

<A>

Anchor Identifies a link or an anchor in a document. Commonly used to create a hyperlink with either a NAME attribute, an HREF attribute, or both.

The HTML page in Listing C.1 presents a link on the screen. When you click on the text "Would you like to see the testing section," the anchor will take you to the text "Testing Section."

LISTING C.1 HTML page with Anchor tags within the same document.

```
<HTML>
<TITLE> This is a test </TITLE>
<BODY>
<A HREF="#Testing"><h2>Would you like to see the testing
section?</h2></A>
<BR><BR><BR><BR><BR><BR><BR><BR><BR><BR><BR><BR>
<BR><BR><BR><BR><BR><BR><BR><BR><BR><BR><BR><BR>
```

```
<A NAME="Testing"><h2> Testing Section </h2>
</BODY>
</HTML>
```

Three optional attributes are used with the LINK:

```
<A NAME="anchor-name">
```

If you have long documents and you want a user to be able to view important sections quickly, you should specify names for sections. The links to the *named section* should be provided wherever the section is being referred, as shown in the example HTML page in Listing C.2, with <A NAME="anchor-name"> and <A HREF="anchor-name"> pair.

```
<A HREF="url">link-text</A>
```

This tag creates a hyperlink to other URL.

LISTING C.2 HTML page with Anchor tag for a hyperlink.

```
<HTML>
<TITLE> This is a test </TITLE>
<BODY>
    <A HREF=http://www.aklinux.com> Find out about Linux </A>
</BODY>
</HTML>
```

```
<A HREF="url#anchor-name">link-text </A>
```

This tag links a section in the other URL:

```
<A HREF=http://www.aklinux.com/index.html#FloatingPoint> Find out about
Floating Point</A>
```

You can use the TARGET attribute as an extension, but it is often used to load a document into a specific frame.

```
TARGET="window name"
<FRAME SRC="frames/list.html" NAME="list" >
<A HREF="/frames/list.html" TARGET="list"> Go to List </A>
```

The preceding code will load the document pointed to by HREF into the window specified by the target "window name." The followings are the predefined target names:

_blank Will cause the link to be loaded into a new blank window.

_self Will cause the link to be loaded into the same window the link is in.

_parent Will cause the link to be loaded into the parent of this document.

_top Will cause the link to be loaded into the full body of this window.

<ADDRESS>

The Address tag shows an address with different fonts. The HTML page in Listing C.3 prints an address in italics.

LISTING C.3 HTML page with Address tag.

```
<HTML><TITLE> This is a test </TITLE><BODY>
My Address is:
<ADDRESS>John Q. Public<BR>Anywhere, NY<BR>00000</ADDRESS>
</BODY></HTML>
```

<APPLET>

The Java applet tag runs a Java applet referred to by a URL. The applet content consists of optional PARAM tags, ordinary text, and markups. See the TEXTFLOW tag also, if no ordinary text and markup is included.

```
Format: <APPLET attributes> applet-content </APPLET>
```

Attributes

CODEBASE="base" What "base" should be used when resolving source-relative URLs.

CODE="code" The URL of the applet to be run.

NAME="applet name" The name of the applet.

ALIGN="alignment" Should be LEFT, RIGHT, TOP, MIDDLE, or BOTTOM.

ALT="text" The text to be displayed by a browser that does not support applets.

HEIGHT=number The height of the applet display area in pixels.

WIDTH=number The width of the applet display area in pixels.

HSPACE=pixels Specifies the horizontal space left and right in which the applet displays.

VSPACE=number The space, in pixels, to leave above and below the applet display area.

DOWNLOAD=number This specifies the order in which applets are downloaded.

HEIGHT=pixels Specifies the height of the applet display area.

NAME=name Identifies an applet to other applets within the HTML page.

TITLE=text Specifies an advisory title string.

<AREA>

The Area tag indicates an area in an image map where a user can choose to link to another document. Valid only within a Map, the tag defines areas that act as hotspots within an image. Typically, a map will have multiple Area tags. If the user clicks on a location that is inside two or more defined areas, the one that appears first within the MAP entry is selected for action. Listing C.4 displays a page where an image is divided into two areas.

LISTING C.4 HTML page showing Map and Area tags dividing an image.

```
<html><head><title>Home Page</title>
</head><body><map name="Advertise">
<Area shape=rect Coords="0,0,150,44"
href="http://www.pakdata.com/urdu">
<Area shape=default href="http://www.cnn.com">
</map>
<img src="KbComp.gif" usemap="#Advertise" width="150" height="89"
alt="KbComp.gif" align="left">
</body></html>
```

Attributes

ALT="text" Provides a textual description for text-only browsers.

SHAPE="{RECT, CIRCLE, POLY}" Attribute can be one of RECT, CIRCLE, POLY, or DEFAULT. See the *Attributes* section for details.

NOHREF Attribute means that clicks here will not cause a link to be followed.

```
<AREA SHAPE="shape" ALT="text" CO-ORDS="co-ords" NOHREF>
```

TARGET If Target is specified, the document pointed to by HREF will be loaded into the window specified by target "window name." See *TARGET Attribute* in Anchor definition

<B>

Boldtext appears.

```
<B> Text </B>   The text appears as Bold
```

<BASEFONT>

BASEFONT Font tag defines the base on which relative FONT changes are based. (Default is 3).

```
<BASEFONT SIZE="10">
```

<BGSOUND>

The Background Sound tag identifies a .wav, .au, or .mid resource that will be played when the page is opened. The optional LOOP attribute will cause the resource to be played n times. The default LOOP="INFINITE" will cause the resource to be played continuously as long as the page is open.

```
<BGSOUND SRC="mysong.wav">
<BGSOUND SRC="URL"LOOP="3">
```

<BIG>

Big shows text in a large size font; the following page displays the text inside the BIG tag with large fonts:

```
<HTML><TITLE> This is a test </TITLE><BODY>
<BIG> Big Font </BIG> Normal Font</BODY></HTML>
```

<BLINK>

Blink highlights the text by having it blink.

```
<BLINK> text </BLINK> Makes text blink on and off.
```

<BQ>
<BLOCKQUOTE>

The Block Quote tag defines text that is made up of more than a few lines. Many browsers (including Netscape) display it in an indented block surrounded by blank lines. Use this tag to display multiple paragraphs in a document. Other suggestions from the HTML 2.0 spec include displaying the text in italics, or starting each line with the Usenet standard quote indicator, > . In HTML 3.0, the CLEAR attribute is used to position a quote after a graphic; it can be LEFT, RIGHT, or ALL, and specifies which margin should be clear. The NOWRAP attribute stops the browser from wrapping except at a BR tag.

Format: <BLOCKQUOTE> text </BLOCKQUOTE>
<BQ> text </BQ>

Attributes

```
<BQ CLEAR = attributes> text </BQ>
<BQ NOWRAP> text </BQ>
```


The Line Break tag breaks the current line of text. It is not necessary inside a PRE element. There is no </BR> tag.

Attributes

```
<BR CLEAR="type">
```

The "type" can be one of the following:

LEFT to break until there is nothing to the left.
RIGHT for the right side.
ALL for break until both sides are clear.
NONE for a normal break.

<CAPTION>

The Caption tag defines the caption of a figure or table. It is valid only within FIG or TABLE tags.
Format: <CAPTION> text </CAPTION>

Attributes

```
<CAPTION ALIGN=alignment> text </CAPTION>
```

The alignment can be LEFT, RIGHT, or CENTER.

```
<CAPTION VALIGN=vertical-alignment> text </CAPTION>
```

The VALIGN attribute arranges for the caption to be at the TOP or BOTTOM of the table or figure.

<CENTER>

The Center tag defines text that should be centered.
Format: <CENTER> text</CENTER>

<COL>

The Table Column tag sets the properties of one table column at a time. Do not use this tag with a COLGROUP element.
Format: <COL> content </COL>

Attributes

<COL ALIGN=alignment> content </COL> The ALIGN attribute specifies the text alignment in the cells within the column. The values for "alignment" are LEFT, MIDDLE, and RIGHT, and the default is MIDDLE.

<COL SPAN=number> content </COL> Indicates the number of columns in the group.

<COLGROUP>

Specifies how a Table Column Group appears on the page. The tag sets the properties of one or more table columns.

Format: <COLGROUP> column data </COLGROUP>

Attributes

<COLGROUP ALIGN="align"> column data </COLGROUP> The values are LEFT, RIGHT, and CENTER (the default).

<COLGROUP VALIGN="valign"> column data </COLGROUP> The VALIGN attribute sets the vertical alignment for the column. The values are TOP, MIDDLE (the default), and BOTTOM.

<COLGROUP HALIGN="halign"> column data </COLGROUP> The HALIGN attribute specifies the horizontal alignment of text in the cells for the column group.

<COLGROUP WIDTH="width"> column data </COLGROUP> The WIDTH attribute specifies the width of each column in the column group.

<COLGROUP SPAN="number"> column data </COLGROUP> The SPAN attribute sets the number of consecutive columns for the group (Listing C.5).

LISTING C.5 HTML page showing columns in a table.

```
<html><head><title>Home Page</title>
</head><body>
<TABLE>
<COLGROUP>
<COL ALIGN="RIGHT">
<COL ALIGN=RIGHT>
<TR>
```

```
<TD> See me at the right </TD>
<TD> See me in the center </TD>
</TR>
</TABLE>
</body></html>
```

<EMBED>

The Embed tag element is used to embed a plug-in into a document. The OBJECT tag can also be used to embed objects.

Format: <EMBED attributes> alternate HTML </EMBED>.

The HTML page in Listing C.6 will request the object mysong.midi.

LISTING C.6 HTML page requesting an embedded object.

```
<HTML><TITLE> This is a test </TITLE><BODY>
<EMBED SRC="mysong.midi">listen to my song</EMBED></BODY></HTML>
```

Attributes

SRC="URL" Identifies the location of the object to be embedded.

HEIGHT=number Specifies the height of the object, according to the UNITS attribute.

WIDTH=number Specifies the width of the object, according to the UNITS attribute.

UNITS=units In *pixels*, meaning the width and height are measured in pixels, or ***en***, meaning the width and height are measured in en spaces.

NAME=text Indicates the name used by other objects or elements to refer to this object.

OPTIONAL PARAMETER=value Specifies any parameters that are specific to the object. Put the name of the parameter in place of "OPTIONAL PARAMETER."

PALETTE=#rgb|#rgb Sets the foreground or background color. The first color is the foreground.

ESCAPE SEQUENCES

The escape sequence allows you to display the reserved characters such as <, > , &, and " in HTML documents. The following HTML page will display the signs (<), (>), (copyright), and (registered trademark).

```
<HTML><TITLE> This is a test </TITLE><BODY>
&lt;&gt;&copy;&reg; </BODY></HTML>
```

&br;
 will display as <BR>.

< Will display as <.

> Will display as > .

& Will display ampersand &.

" Will display quotes.

** ** Will display spaces inside text.

® Will display registered trademark.

© Will display copyright symbol.

<FORM>

Web forms are the most popular method of interactive HTML pages, interfacing with users. Listing C.7 is a Web form that displays a text window and a Submit button, and requests text input of 10 characters and the window title Name.

LISTING C.7 HTML page shown on the right column.

```
<HTML>
<HEAD><TITLE> Right column</TITLE></HEAD>
    <body>
<FORM ACTION="http://localhost/security.cgi" method="post">
    Name: 
    <INPUT TYPE=text size=10 name=Name>
    <INPUT TYPE=submit value="Enter">
</FORM>
</HTML>
```

Attributes

CHARSET="..." Specifies the character encoding for input data. The following line shows an attribute with the FORM tag indicating character set ISO-8859-1:

```
<FORM METHOD =POST ACCEPT-CHARSET="ISO-8859-1" ACTION="cgi-
bin/form.cgi>
```

ACCEPT="..." Specifies a list of mime types recognized by the server.

CLASS="..." Specifies the style class applied to the form.

METHOD="POST|GET" Specifies how the data should be returned to the server.

<FRAME>

Frame tags help you divide a window into sections, thus allowing you to see more than one document in the HTML page. The tag appears inside the FRAMESET tag, and specifies one frame in the frameset. In order to create a FRAMESET document, you need to declare the following line as the first line of your document:

```
<!DOCTYPE HTML PUBLIC "-//W3C//DTD HTML 4.0 Frameset//EN">
```

The example in Listing C.8 shows a page with two columns.

LISTING C.8 HTML page with Address tag.

```
<!DOCTYPE HTML PUBLIC "-//W3C//DTD HTML 4.0 Frameset//EN">
<HTML><HEAD><TITLE> Two columns</TITLE></HEAD>
<FRAMESET COLS="1*,3*">
   <FRAME SRC="Left.html">
   <FRAME SRC="Right.html">
</FRAMESET>
</HTML>
```

The COL="1*,3*" attribute specifies that the window should be divided into it two columns; column 1 is one-third of column 2. Column 1 should be loaded with Left.html, while column 2 should be loaded with Right.html page. Listing C.9 is the left.html page, and Listing C.10 is the right.html page.

LISTING C.9 HTML page shown on the left column.

```
<HTML>
<HEAD><TITLE> Left column</TITLE></HEAD>
    <body> <h1> Left Column </h1></body>
    </HTML>
```

LISTING C.10 HTML page shown on the right column.

```
<HTML>
<HEAD><TITLE> Right column</TITLE></HEAD>
    <body> <h1> Right Column </h1></body>
    </HTML>
```

Attributes

SRC="URL" The URL of the source document to be displayed in this frame. If the frame does not specify a source, it will be displayed as blank space.

NAME="window name" The name associated with the frame. The TARGET attribute in the A, BASE, AREA, and FORM tags can use this attribute to target this frame.

MARGINWIDTH=number The number is the width in pixels.

MARGINHEIGHT=number The number is the top and bottom margin thickness in pixels.

SCROLLING=type Here, type is one of yes, no, or auto. It specifies if the frame is to have a scroll bar; auto (the default) means the browser will decide.

NORESIZE Stops the user from resizing the frame.

FRAMEBORDER=yes|no Specifies if the border should be displayed.

FRAMESPACING=number Here, number is the spacing between frames in pixels.

ALIGN = "alignment" One of LEFT, RIGHT, TOP, TEXTTOP, MIDDLE, BSMIDDLE, BASELINE, BOTTOM, or ABSBOTTOM.

<FRAMESET>

The Frameset tag replaces the Body tag in a document, and is used to split the document's window into a set of smaller frames. Please see the <FRAME> tag and Listing C.6 for <FRAMESET> tag use. The Frameset can be nested to create more complicated frame layouts. Noframe tags can also be placed in a frameset.

Format: <FRAMESET attributes> frame tags </FRAMESET>

Attributes

ROWS="row heights" Specifies a list of values for the rows; each can be specified as a percentage, a pixel value, or as "*". The frameset will be split vertically into frames based on these values. Rows with "*"'s in them will have remaining space split between them.

COLS="column widths" Specifies a list of values for the columns. The width of each column can be specified as a percentage, a pixel value, or as "*". The frameset will be split into frames based on these values. Columns with a width of "*" will split the space that is not assigned to other columns.

<H1>

The Heading 1 tag defines a level-1 heading. It is typically shown in a very large bold font with several blank lines around it, and is used by automatic indexers to describe a page.

Format: <H1> text </H1>

Attributes

```
<H1 ALIGN=alignment > text </H1>
```

The alignment attribute can be LEFT, RIGHT, or CENTER; it defines the placement of the header on the screen.

<H1 SRC="URL" > text </H1>

<H1 DINGBAT="entity-name" > text </H1>

<H1 NOWRAP> text </H1>

<H1 CLEAR=clear > text </H1>

The SRC attribute identifies a graphic image to be embedded before the header text, while the DINGBAT attribute identifies an iconic entity to be embedded there. The Clear attribute is used to position a header after a graphic; it can be LEFT, RIGHT, or ALL, and specifies which margin should be clear. The NOWRAP attribute prevents the browser from breaking long header lines; use a BR tag to break those lines yourself.

<H2>

The Heading 2 tag defines a level-2 heading. It is typically shown in a large bold font with several blank lines around it.

Format: <H2> text </H2>
Attributes: Please see the attributes of H1 tag.

<H3>

The Heading 3 tag defines a level-3 heading. It is typically shown in a large italic font, slightly indented, with blank lines around it.

Format: <H3> text </H3>
Attributes: Please see the attributes of H1 tag.

<H4>

The Heading 4 tag defines a level-4 heading. It is typically shown in a bold font, indented more than a level-3 heading, with blank lines around it.

Format: <H4> text </H4>
Attributes: Please see the attributes of H1 tag.

<H5>

The Heading 5 tag defines a level-5 heading. It is typically shown in an italic font, indented the same as a level-4 heading, with a blank line above it.

Format: <H5> text </H5>
Attributes: Please see the attributes of H1 tag.

<H6>

The Heading 6 tag defines a level-6 heading. It is typically shown in a normal font, indented more than a level-5 heading, with a blank line above it.

Format: <H6> text </H6>
Attributes: Please see the attributes of H1 tag.

<HR>

The Horizontal Rule tag causes a horizontal line to be drawn across the screen. There is no </HR> tag.

Attributes

<HR SRC= "URL">

The SRC attribute specifies an image file to be used for the rule.

<HR SIZE= number>

The SIZE attribute allows you to specify the thickness of the line (in pixels) with the SIZE attribute.

<HR WIDTH= number%>

The WIDTH attribute governs what percentage of the screen width is occupied by the rule.

<HR ALIGN= alignment%>

The ALIGN attribute aligns a rule that is smaller than the screen: alignment can be LEFT, RIGHT, or CENTER.

<HR NOSHADE>

The NOSHADE attribute prevents the browser from using any shading or three-dimensional effects.

<HR COLOR="colorname">

See the *Attribute* section for details of colornames.

<HR COLOR="#rrggbb">

See the *Attribute* section for details of color codes.

<IMG>

Inline Image allows you to add graphic images to your pages. The example in Listing C.11 inserts a picture in the current page.

LISTING C.11 HTML page shown on the right column.

```
    <HTML>
    <HEAD><TITLE> My image</TITLE></HEAD>
<body><P><IMG SRC="//www.aklinux.com/images/sunset.jpg" ALT="sunsets in
the west" ALIGN=TOP></P>
    </body>
    </HTML>
```

Attributes

ALIGN="alignment"

Should be one of LEFT, RIGHT, TOP, TEXTTOP, MIDDLE, ABSMIDDLE, BASELINE, BOTTOM, or ABSBOTTOM. This causes the top, middle, or bottom of the image to be aligned with the text. The LEFT or RIGHT moves the image to the left or right of the screen and allows text to flow around it.

ALT="text"

The text to be displayed by a browser that does not display images, such as Lynx, or to be used when image display is suppressed.

BORDER=number

The border thickness in pixels. (Do not use BORDER=0 for images that are links.)

CONTROLS

If present, and an AVI resource being played, displays controls under the resource.

DYNSRC="url"

Specifies an AVI resource to be played, or a VRML world. Always include a still image as well with the SRC attribute, for use by browsers that do not display inline video or VRML.

HEIGHT=number Specifies the height of the image, according to the UNITS attribute.

HSPACE=number The space, in pixels, to leave to the left and right of the image.

ISMAP="url" Indicates that this image is a server-side image map.

LOOP=number The optional LOOP attribute will cause the resource to be played a certain number of times. LOOP="INFINITE" will cause the resource to be played continuously as long as the page is open.

LOWSRC="url" Specifies an image to be displayed while the SRC image is being loaded. This alternate image should take much less time to download than the SRC image, and should be lower resolution: for example, black and white.

START="start option" The START attribute specifies when the browser should start to play the resource specified with the DYNSRC attribute. START=FILEOPEN instructs the browser to play the resource only when the file is opened. START=MOUSEOVER instructs the browser to play the resource each time the user moves the mouse cursor over it. START=FILEOPEN, MOUSEOVER does both.

SRC="URL" Identifies the image source; typically, a GIF or JPEG file.

UNITS=units Either pixels, meaning the width and height are measured in pixels, or **en**, meaning the width and height are measured in en spaces.

USEMAP="url" Overrides the ISMAP attribute, if present, and if the browser supports client-side image maps. It uses the MAP element found at the URL to translate clicks.

WIDTH=number Specifies the width of the image, according to the UNITS attribute.

VSPACE=number The space, in pixels, to leave above and below the image.

<INPUT>

The Input tag is used mainly in the form tags for specifying variables that are updated by the users. Listing C.12 illustrates a form that requests a name and address from the user.

LISTING C.12 A Web form with input values.

```
<HTML><head><title>Home Page</title>
</head><body>
    <FORM ACTION="/cgi-bin/customer/" METHOD=POST>
    <INPUT NAME="name" TYPE="TEXT" SIZE=5>
    </FORM>
    </body>
    </HTML>
```

Attributes

<INPUT TYPE=file> File-upload fields enable the user to specify files for uploading to the server. (The form must be defined as multipart/form-data.)

Format: <INPUT TYPE=file Name=MyFile>

<MAP>

The Map tag defines an image map that divides an image into sections, where each section can point to a different URL. The tag gives a name to a collection of Area tags that connect the user clicks with URLs. Image maps allow the user to interact with the page by selecting a part of the image to navigate to another page. Image maps are simply files that contain a list of area shapes on an image and their corresponding URLs. When a pixel on the image is clicked, the URL associated with the area containing that pixel is returned.

Format: <MAP NAME="name"> area tags </MAP>

See Listing C.4 for an HTML page that displays an image divided into two Area maps.

Attributes

The following are the pixel area groupings:

circle Define the center and edge point.

rect Define the upper-left and lower-right points.

point Define a single point.

poly Define a polygon of points given n number of points.

default Rest of the points.

<A HREF="images/simple.map">

Read the mapping attributes from the file simle.map. The image map is stored on the Web server and is referred to using a standard URL. Following is an example of a simple image map, simple.map:

```
circle logo.html 20,20 30,30
rect acct.html 20,20,40,40
```

<IMG SRC="images/logo.gif" ISMAP> </A>

You can use these image maps together with the <A> and <IMG> HTML tags. Assume the simple.map described earlier is in an images directory, and the image file is logo.gif.

<MARQUEE>

Marquee displays scrolling text message within a document as shown in the example HTML page in Listing C.13 file MARQUEE.html. The tag defines a moving piece of text.

LISTING C.13 HTML page showing scrolling text.

```
<HTML><head><title>Home Page</title>
</head><body>
<MARQUEE DIRECTION=LEFT BEHAVIOR=SCROLL SCROLLDELAY=250
SCROLLAMOUNT=10> Marathon Sale going on </MARQUEE>
</body>
    </HTML>
```

Format

<MARQUEE> text </MARQUEE>

Attributes

<MARQUEE ALIGN="align"> text </MARQUEE> The ALIGN attribute works like the ALIGN attribute in the IMG tag, setting the location of the surrounding text. "align" can be TOP, BOTTOM, or MIDDLE.

<MARQUEE BEHAVIOR="behavior"> text </MARQUEE> The BEHAVIOR attribute defines the way the text moves. SCROLL means that the text slides into the marquee box and out again, and then repeats. SLIDE means that the text slides into the marquee box, stops when it is all in, and then repeats. ALTERNATE means that the text bounces back and forth within the marquee box.

<MARQUEE BGCOLOR="#rrggbb"> text </MARQUEE> The BGCOLOR attribute specifies the color to be used for the background. See the *Attribute* definition for color codes rrggbb.

<MARQUEE BGCOLOR="colorname"> text </MARQUEE> Color names can be used for the BGCOLOR attribute. Valid colors are Aqua, Black, Blue,

Fuchsia, Gray, Green, Lime, Maroon, Navy, Olive, Purple, Red, Silver, Teal, Yellow, and White.

<MARQUEE DIRECTION="direction"> text </MARQUEE> The DIRECTION attribute is LEFT or RIGHT and specifies the direction in which the text should move.

<MARQUEE HEIGHT=n> text </MARQUEE>

<MARQUEE HEIGHT=n%> text </MARQUEE>

<MARQUEE WIDTH=n> text </MARQUEE>

<MARQUEE VSPACE=n> text </MARQUEE>

<MARQUEE WIDTH=n%> text </MARQUEE>

The HEIGHT and WIDTH attributes size the marquee box. If n is an absolute number, it is taken to mean pixels; when n is followed by a % sign, it means a percentage of the width or height (as appropriate) of the screen.

<MARQUEE HSPACE=n> text </MARQUEE> The HSPACE and VSPACE attributes specify a margin to the left and right, or above and below the marquee box, in pixels.

<MARQUEE LOOP=n> text </MARQUEE> The LOOP attribute will cause the marquee to scroll n times. LOOP="INFINITE" will cause the marquee to scroll as long as the page is open.

<MARQUEE SCROLLAMOUNT=n> text </MARQUEE>

<MARQUEE SCROLLDELAY=n> text </MARQUEE>

The SCROLLAMOUNT attribute specifies the amount, in pixels, to move the scrolling text each time it is drawn. The SCROLLDELAY attribute specifies the delay, in milliseconds, between drawings.

<MULTICOL>

The Multi Column Text tag is a container used to split the display into columns without using frames or tables.

Attributes

COLS="number" The COLS attribute is mandatory, and controls how many columns the display will be split into.

GUTTER="pixels" The GUTTER attribute controls the pixels of space between columns. It defaults to a value of 10.

WIDTH="number" The WIDTH attribute controls the width of an individual column.

<NOBR>

The No Break tag defines a block of text that will have no line breaks, except those explicitly requested with BR or suggested with WOBR.

Format: <NOBR> text </NOBR>

<NOFRAMES>

The Noframes tag specifies HTML that can be used by browsers that do not support frames. Everything between the start and end tags is ignored by browsers that understand frames.

Format: <NOFRAMES> alternate HTML </NOFRAMES>

<OBJECT>

The Object tag is similar to the Embed tag, with extended parameters. The Object can be an ActiveX object, a QuickTime movie, or any other object that the browser supports.

Format: <OBJECT> object-content </OBJECT>

<P>

The Paragraph tag starts a new paragraph, equivalent to two BR tags. The </P> tag is optional, but if the tag is present, it is only for inserting space between two paragraphs, but necessary when attributes (for example, ALIGN="center") are to apply to the whole paragraph. The ALIGN attribute can be one of LEFT, RIGHT, or CENTER. The NOWRAP attribute will make it so lines are only broken at the <BR> tag.

Format: <P> text </P>

Attributes

<P ALIGN={LEFT,CENTER,RIGHT,JUSTIFY}> text </P> Aligns paragraph text flush left, flush right, or in the center of the documents.

<P NOWRAP> text</P> No Wrap specifies the text will not wrap around when maximum line length is reached.

<PARAM>

This tag is valid only within an Applet or Object tag. It passes parameters to the applet, which calls the getParameter() to get the actual parameters.

Format: <PARAM NAME="name" VALUE="value">

<PLAINTEXT>

The Plaintext tag defines text that should be shown in a fixed width font with the line breaks and other whitespace characters specified by the page. There is no need to use <BR> tags to indicate line breaks. The line breaks in the source text will be taken care of by the browser. In addition, multiple spaces will be displayed as multiple spaces.

Format: <PLAINTEXT> text </PLAINTEXT>

<PRE>

The Preformatted text tag is obsolete.

<SCRIPT>

The Script tag identifies script code. The script code to be executed at this point of the document, or may contain functions for use later in the document. Netscape Navigator supports JavaScript, and Microsoft Internet Explorer 3.0 supports both JavaScript and VBScript. The statements are usually enclosed in the comment tag, so that browsers that do not support scripting do not render the code as text. Functions used by the document are usually defined in the Head tag so that they are loaded and available before the user does anything that might call them.

Format: <SCRIPT LANGUAGE="language">
<!— script statements —></SCRIPT>

<SPACER>

The Spacer tag creates an area of whitespace within the document.

Attributes

TYPE=type

The TYPE attribute has three possible values: horizontal (the default,) vertical, and block.

SIZE=pixels

The SIZE attribute only applies when the spacer has a type of horizontal or vertical. The attribute controls the absolute width or height in pixels of the spacing added.

WIDTH=pixels

The WIDTH attribute only applies when the spacer is of type block. The attribute controls the absolute width in pixels of the spacing rectangle added.

HEIGHT=pixels

The HEIGHT attribute only applies when the spacer is of type block. The attribute controls the absolute height in pixels of the spacing rectangle added.

ALIGN=alignment

The ALIGN attribute only applies when the spacer is of type block. The attribute controls the alignment of the spacing rectangle in exactly the same way it would control the alignment of an IMG tag. "alignment" should be one of LEFT, RIGHT, TOP, TEXTTOP, MIDDLE, ABSMIDDLE, BASELINE, BOTTOM, or ABSBOTTOM.

<WBR>

The Word Break tag identifies a place where a word can be broken, or where a line can be broken within a NOBR block.

<TABLE>

A table consists of an optional caption (CAPTION) and one or more rows (TR).

Format: <TABLE attributes> table-content </TABLE>

Attributes

ALIGN="alignment"

This causes the table to be aligned in one of a variety of ways on the page. Here, "alignment" should be one of:

LEFT: To the left text margin.

CENTER: In the center of the page (turns on NOFLOW).

RIGHT: To the right text margin.

BLEEDRIGHT: To the rightmost edge of the window.

BLEEDLEFT: To the leftmost edge of the window.

JUSTIFY: To both text margins (the table size is adjusted; turns on NOFLOW).

Note that this does not affect the alignment of the table entries.

WIDTH=number

The UNITS attribute is used to translate number.

BORDER Causes the table to be drawn with a border.

BORDER=number Draws the table with a border number of pixels thick.

CELLPADDING=number Separates the cell borders and the text with a padding of number pixels.

CELLSPACING=number Separates cells with a gutter of number pixels.

BGCOLOR="#rrggbb" See the *Attributes* section for details of color codes.

BGCOLOR="colorname" Sets the background color for the entire table.

BORDERCOLOR="#rrggbb" See the *Attributes* section for details of color codes.

BORDERCOLOR="colorname" Sets the border color for the entire table.

BORDERCOLORLIGHT="#rrggbb" See the *Attributes* section for details of color codes.

BORDERCOLORLIGHT="colorname" Sets the border highlight color for the entire table.

BORDERCOLORDARK="#rrggbb" See the *Attributes* section for details of color codes.

BORDERCOLORDARK="colorname" Sets the border shadow color for the entire table.

VALIGN="valign" Sets the vertical alignment for the entire table. "valign" is TOP or BOTTOM.

CLEAR=clear Should be one of LEFT, RIGHT, or ALL, and specifies which margin should be clear.

NOFLOW Prevents text flow around the table and is equivalent to setting the CLEAR attribute on the element after the table.

COLSPEC=colspec A list of column alignments and widths, separated by spaces. There should be one entry for each column in the table, and each should be an optional capital letter for alignment: one of L (left), R (right), C (center), J (justify), or D (decimal) followed immediately by a number describing the width.

UNITS=units This makes sense only if the COLSPEC or WIDTH attributes are being used, and specifies the units to be used for the column or table widths. Units should be one of the following:

> **relative** Column widths are a percentage of the entire table width; table width is a percentage of the entire window width.
>
> **DP="character"** The character (the default is ".") to be used in aligning for decimal point alignment.
>
> **NOWRAP** Prevents word wrap within table entries.

<TBODY>

The Table Body tag is used to group together a number of rows within a table, for assigning ID or STYLE values.

Format: <TBODY> table body </TBODY>

Attributes

CLASS=type Indicates the class to which the element belongs.

ID=value Specifies a unique value for the element over the document.

STYLE=css1 properties Specifies the style information.

<TD>

The Table Data tag is valid only within the TR tag. The tag defines a table cell.

Attributes

ALIGN="alignment" Causes the table to be aligned in one of a variety of ways on the page. Here, "alignment" should be one of:

LEFT: To the left text margin.

CENTER: In the center of the page (turns on NOFLOW).

RIGHT: To the right text margin.

JUSTIFY: To both text margins (the table size is adjusted; turns on NOFLOW).

Note that this does not affect the alignment of the table entries.

CHAR: Aligns on a specific character.

WIDTH=number The UNITS attribute is used to translate numbers.

BORDER Causes the table to be drawn with a border.

BORDER=number Draws the table with a border number pixels thick.

CELLPADDING=number Separates the cell borders and the text with a padding of number pixels.

CELLSPACING=number Separates cells with a gutter of number pixels.

BGCOLOR="#rrggbb" or "..." See the *Attributes* section for details of color codes.

BGCOLOR="colorname" Sets the background color for the entire table.

BORDERCOLOR="#rrggbb" See BGCOLOR for color values.

BORDERCOLOR="colorname" Sets the border color for the entire table.

BORDERCOLORLIGHT="#rrggbb" See BGCOLOR for color values.

BORDERCOLORLIGHT="colorname" Sets the border highlight color for the entire table.

BORDERCOLORDARK="#rrggbb" See BGCOLOR for color values.

BORDERCOLORDARK="colorname" Sets the border shadow color for the entire table.

VALIGN="valign" Sets the vertical alignment for the entire table. "valign" is TOP or BOTTOM.

CLEAR=clear Should be one of LEFT, RIGHT, or ALL, and specifies which margin should be clear.

NOFLOW Prevents text flow around the table and is equivalent to setting the CLEAR attribute on the element after the table.

COLSPEC=colspec A list of column alignments and widths, separated by spaces. There should be one entry for each column in the table, and each should be an optional capital letter for alignment: one of L (left), R (right), C (center), J (justify), or D (decimal) followed immediately by a number describing the width.

UNITS=units Defined only if the COLSPEC or WIDTH attributes are being used, and specifies the units to be used for the column or table widths. Units should be one of the following:

relative Column widths are a percentage of the entire table width; table width is a percentage of the entire window width.

DP="character" The character (the default is ".") to be used in aligning for decimal point alignment.

NOWRAP Prevents word wrap within table entries.

<TEXTAREA>

The Textarea tag defines a multiple-line text input field within a form.

Format: <TEXTAREA NAME="message"> text </TEXTAREA>

The HTML page in Listing C.14 opens a text box to be filled in by the user.

LISTING C.14 HTML page showing a text box.

```
<HTML><head><title>Home Page</title>
</head><body>
    <TEXTAREA NAME="message" COLS=50 ROWS=5> Greetings </TEXTAREA>
    </body>
    </HTML>
```

<TFOOT>

Defines the table footer. The Footer tag is used to group all footers.

Attributes

ALIGN=left, center, right, justify Specifies the alignment of text in the heading; the default is center.

CLASS=type Indicates the style class to which the element belongs; for example:

<THEAD CLASS="casual">

ID=value Specifies a unique value for the element over the document. When a style is assigned to that ID selector, it affects only the particular tag that matches the ID.

STYLE=css1properties Specifies the style of the information.

VALIGN=middle/top/bottom Specifies the vertical alignment of text in the heading; the default is middle.

<TH>
<THEAD>

The Thead tag defines the table heading. The Thead tag is used to group all headers together.

Attributes

Valid only in a TR, the table header tag defines a header cell.

COLSPAN="number" The number of columns this header occupies.

ROWSPAN="number" The number of rows this header occupies.

NOWRAP This attribute prevents word wrap within the cell.

ALIGN="align" Governs the alignment of the text within the table cell. "align" can be LEFT, RIGHT, or CENTER. Table Header cells default to centered.

VALIGN="align" Governs the alignment of the text within the table cell. "align" can be TOP, MIDDLE, BOTTOM, or BASELINE.

BGCOLOR="#rrggbb" See the *Attributes* section for details of color codes.

BGCOLOR="colorname" See the *Attributes* section for details of color names.

These attributes set the background color for the header cell:

BORDERCOLOR="#rrggbb" See the *Attributes* section for details of color codes.

BORDERCOLOR="colorname" See the *Attributes* section for details of color names. These attributes set the border color for the header cell.

BORDERCOLORLIGHT="#rrggbb" See the *Attributes* section for details of color codes.

BORDERCOLORLIGHT="colorname" See the *Attributes* section for details of color names. These attributes set the border highlight color for the header cell.

BORDERCOLORDARK="#rrggbb" See the *Attributes* section for details of color codes.

BORDERCOLORDARK="colorname" See the *Attributes* section for details of color codes. These attributes set the border shadow color for the header cell.

<TR>

The Table Row tag is valid only in a Table. The tag defines a row of cells that are defined with TH and TD tags.

Attributes

ALIGN="align" Governs the alignment of the text within the table cell. "align" can be LEFT, RIGHT, or CENTER.

VALIGN="align" Governs the alignment of the text within the table cell. "align" can be TOP, MIDDLE, BOTTOM, or BASELINE.

BGCOLOR="#rrggbb" See the *Attributes* section for details of color codes.

BGCOLOR="colorname" See the *Attributes* section for details of color names. This attribute sets the background color for the table row.

BORDERCOLOR="#rrggbb" See *Attributes* section for details of color codes.

BORDERCOLOR="colorname" See the *Attributes* section for details of color names. This attribute sets the border color for the table row.

BORDERCOLORLIGHT="#rrggbb" See the *Attributes* section for details of color codes.

BORDERCOLORLIGHT="colorname" See the *Attributes* section for details of color names. This attribute sets the border highlight color for the table row.

BORDERCOLORDARK="#rrggbb" See the *Attributes* section for details of color codes.

BORDERCOLORDARK="colorname" See the *Attributes* section for details of color names. This attribute sets the border shadow color for the table row.

CLASS="class" One of Header, Body, or Footer, and allows the browser to arrange for header or footer rows to be displayed as the user scrolls through the document.

<SELECT>

Used inside the Option tag and specifies a selection list similar to listboxes, except they return a single value (Use SIZE=1).

```
<SELECT NAME='State' SIZE=1>
<OPTION> CA
<OPTION> WA
<OPTION> OR
<OPTION> NV
```

<TEXTFLOW>

If a Java applet element has no ordinary text and markup in its applet content, and contains only PARAM tags, comments, and spaces, you should add a Textflow tag for SGML parsers.

SUMMARY

In this appendix, we defined the basic construct of the Web pages from the markup language point of view, using HTML tags and HTML attributes. We presented a comprehensive list of tags and attributes with examples to show the effect of each tag as it appears when viewed through a Web browser such as Netscape.

Appendix

D About the CD-ROM

The CD-ROM that accompanies *Practical Linux Programming: Device Drivers, Embedded Systems, and the Internet* includes complete source code for all examples presented in the book, grouped by chapter. These examples describe how to:

- Create embedded Linux, a minimum Linux system that can be placed on less than 2 Meg disk space. Scripts are provided that can be modified and customized to create a boot disk and a root file system disk of your choice.
- Setup a virtual Web hosting system to be used as the user interface for your embedded systems, complete with the DNS server, SQL server, and the HTTP server setup.
- Design device drivers that will form the backbone of your embedded system's hardware interface. It is an essential part of all embedded systems that requires an external hardware.
- Interconnect and link your software modules with the interprocess communication facilities of the Linux kernel. The mechanisms of semaphores, shared memories, and message queues are described, creating glue logic for modules in your embedded systems.
- Create dynamic Web pages using Perl scripts, SQL queries, and HTML tags for the user interface part of your embedded systems.
- Create products with the design philosophy of user interface, system interface, and database.

THIRD PARTY SOFTWARE

The CD-ROM also contains demo versions of third party software, including a Web database management program (Dodobase 2.0) a UNIX cross-platform printing system (CUPS and ESP Print Pro 4.2.1), and an Integrated Development Environment for C and C++ on GNU/Linux (Anjuta).

Dodobase 2.0 (trial demo)

Web Sight Technologies

http://www.websight-tech.com

Dodobase is a Web database management program that allows Web site owners to create and manage content instantly. Conventional CGI based database programs are slow, take a heavy load on the server, and pose a security risk. Dodobase uses Java Servlets and is fast, secure, and dependable. It can be fully customized to suit the appearance of your Web site and connects to any RDBMS like MS-SQL, MYSQL, and Oracle.

The demo version of Dodobase 2.0 provided on the CD-ROM is a 30-day free trial.

CUPS (Common UNIX Printing System)
ESP Print Pro 4.2.1 (trial demo)

Easy Software Products

http://www.easysw.com

ESP Print Pro is a complete UNIX cross-platform printing system. ESP Print Pro enhances the printing system that comes with your operating system with the Common UNIX Printing System, or "CUPS." CUPS is based on an international printing standard called the Internet Printing Protocol, or "IPP." IPP provides the framework for modern printing and is used by Microsoft, Hewlett-Packard, Xerox, and many other server and printer manufacturers. What you end up with is a more functional, secure, and cross-platform UNIX printing system. Instead of contending with several printing interfaces, there is now only one standard for all systems.

The demo version of ESP Print Pro requires a software license from Easy Software Products to function. You need to contact Easy Software Products to obtain a fully-functional 21-day free trial license before using the software. The trial period begins upon receipt of a license from Easy Software Products.

Anjuta

Naba Kumar

http://anjuta.sourceforge.net

Anjuta is a versatile Integrated Development Environment (IDE) for C and C++ on GNU/Linux. It has been written for GTK/GNOME, and features a number of advanced programming facilities. These include project management, application wizards, an on-board interactive debugger, and a powerful source editor with source browsing and syntax highlighting.

Anjuta is an effort to marry the flexibility and power of text-based command-line tools with the ease-of-use of the GNOME graphical user interface. That is why it has been made as user-friendly as possible.

System Requirements

Red Hat Linux 7.1 or higher
IBM PC compatible system
Pentium II or better
256 MB RAM
10/100 Mbps Ethernet card
2.5 MB free hard disk space
CD-ROM

Please switch the parallel port mode to SPP using the BIOS CMOS setup before running the example software.

Installation

The examples provided with the book CD require Linux Kernel version 2.4 or above installed. A custom-class Red Hat Linux installation is used with the following components selected:

- X Window System
- GNOME
- DOS/Windows Connectivity
- Networked Workstation
- SQL Server
- Web Server
- DNS Name Server

- Emacs
- Development
- Kernel Development
- Utilities

Copy the contents of the "Book" folder to your local hard drive using the following command before compiling and executing the test software in each chapter:

```
#mount -t iso9660 /dev/cdrom /mnt/cdrom
#mkdir /book
#cp -R /mnt/cdrom /book
```

Glossary

Address Resolution Protocol (ARP) Determines the MAC layer address associated with a logical network address.

Application layer Layer 7, or the topmost layer of the OSI model. This layer exposes all the network services to the applications. When an application accesses the network, all actions are carried out through this layer.

Application server An application server runs all or part of an application on behalf of the client, and then transmits the result to the client for further processing.

Asynchronous Transfer Mode (ATM) A cell-based networking technology that transmits data at high speeds.

Attachment Unit Interface (AUI) The connector used with Thicknet.

attenuation A measure of how much a signal weakens as it travels through a medium.

bandwidth The measure of the capacity of a medium to transmit data.

baseline A collection of performance statistics depicting the average (typical) behavior of the network. Later fluctuations in network performance can be measured against the baseline.

beaconing A process used by Token Ring networks to narrow the portion of the ring in which a problem is most likely to exist.

bridge Passes only frames targeted for a computer on the other side of the bridge (and all broadcast frames).

broadband Two or more communication channels can share the bandwidth of the communications medium.

broadcast storm A sudden deluge of network traffic often caused by a faulty network adapter.

brouter A device that can simultaneously perform both routing and bridging.

bus A topology in which all devices connect to a common shared cable.

carrier detection Transmission method in which computers continue to listen to the network as they transmit, in order to detect whether another signal interferes with their signal.

Carrier Sense Multiple Access with Collision Avoidance (CSMA/CA) Each computer signals a warning that says it is about to transmit data. The other computers then wait for the transmission.

Carrier Sense Multiple Access with Collision Detection (CSMA/CD) Carrier detection and carrier sensing used together form the protocol.

carrier sensing Transmission method in which computers listen to see if the network is busy before they attempt to transmit.

Category 5 Data-grade cable, which consists of four twisted pairs and can support data rates of 100 Mbps.

clients The computers that use the shared resources in a server-based network.

CRC Cyclical redundancy check. A procedure used on disk drives to verify that data written to a sector is read correctly later, and to check for errors in data transmission.

CSLIP A compressed version of SLIP.

Data Link Control (DLC) Protocol most commonly used to access mainframes and Hewlett-Packard JetDirect network printers.

Data Link layer The Data Link layer of the OSI model adds information to each packet coming from the Network layer.

datagram An independent data packet being transported by a stateless protocol.

digital volt meter (DVM) A hand-held electronic measuring tool that checks the voltage of network cables.

disk duplexing Disk duplication using a separate controller for each disk.

disk mirroring Duplication of data across multiple disks, which means that any failed disk can simply be replaced.

domain model In a domain model, one security account manager (SANI) database is maintained for all members of the domain in a network.

Domain Name System (DNS) Provides a name as a service to client applications.

event log A log of system events. Windows NT's event log includes a system log, a security log, and an application log. You can view the event log using Windows NT's Event Viewer tool.

fault-tolerant The capability of a system to withstand a failure and continue to function.

fiber-optic cable A glass or plastic tube designed with the principle of total internal reflection to help transmit light.

file server A computer whose main task is to provide file sharing to computers on a network.

File Transfer Protocol (PTP) Enables users to transfer files between diverse host types.

frame relay A type of WAN packet service that is especially well-suited for data traffic.

frequency-division multiplexing (FDM) Dividing bandwidth into frequency bands on broadband media.

frequency hopping Switching among several available frequencies, yet staying on each frequency for a specified interval of time.

gateway A device or program that enables communication between systems that use dissimilar protocols.

hub A central connection point where all workstations are physically connected. A multiport repeater.

Integrated Services Digital Network (ISDN) A set of standards for delivering digital telephone service.

Internet Protocol (IP) A connectionless protocol that provides datagram service.

internetwork A set of connected networks.

Internetwork Packet exchange protocol/Sequenced Packet exchange (IPX/SPX) A protocol suite used by NetWare and Microsoft NWLink stack.

IP address A logical network and host ID that consists of a 32-bit number. It is typically represented in a four-octet dotted-decimal form.

late collision When a collision is missed on an 802.3 network due to propagation delay.

local area network (LAN) A group of computers interconnected within a building or campus setting.

Medium Access Control address (MAC address) The NIC address on an 802.5 network. Each workstation has a unique MAC address.

NetBIOS The Network Basic Input/Output System, which extends all the way to the Session layer and is used to handle naming and file services.

network A group of interconnected computers that share information and resources.

network adapter A device that provides physical access to a network. The term *network adapter* is typically used for an internal card that connects a PC to an Ethernet or Token Ring network.

Network Device Interface Specification (NDIS) Microsoft and 3Com jointly developed the Network Device Interface Specification in 1989. This standard defined an interface between the MAC sublayer and higher layers of the OSI model.

Network File System (NFS) A family of file-sharing protocols for TCP/IP.

Network layer The Network layer of the OSI model determines the route a packet must take to reach its destination.

Network Monitor A software-based tool that monitors network traffic, displaying packet information and keeping statistics on network usage.

New Technology File System (NTFS) The transaction-based file system native to Windows NT.

NWLink Microsoft's version of the IPX/SPX protocol.

ODI Open Data-Link Interface was jointly developed by Novell and Apple Computer Corporation. The goals of ODI are similar to the goals of NDIS: to provide a seamless integration between the Data Link layer and the Transport layer of the protocol stack.

OSI model A seven-layer model used primarily as a reference model.

packet A unit of network data transmitted from a sending PC to a receiving PC.

peer-to-peer network The simplest network configuration that allows users to access each other's resources.

Performance Monitor A tool included with Windows NT that monitors system and network statistics.

Physical layer The lowest layer of the OSI model.

Plenum-grade cable Cabling specially designed to be used without conduit in areas where the fire code prevents PVC cabling.

Point-to-Point Protocol (PPP) A very popular dial-up protocol that allows dynamic negotiation of IP addresses and multiple protocols over a single serial connection.

Presentation layer The Presentation layer defines the format used by applications to exchange data.

protocol analyzer A hardware or combined hardware and software product that monitors network traffic, tracks network performance, and analyzes packets.

protocol binding A potential logical pathway from a network transport protocol (such as TCP/ IP) to a specific adapter. In Windows NT, you must also bind the protocol to certain higher-level network services, such as the server and workstation services.

Protocol Manager (PROTMAN) The Protocol Manager routes packets from the MAC layer to the correct protocol stack in the NDIS model.

redirector A software module (typically implemented as a file system driver) that redirects I/O requests to the network, and thus allows a networked computer to act as a client.

Redundant Arrays of Inexpensive Disks (RAID) A method of paralleling many disks to make them appear as one large disk with the option of fault tolerance.

Remote Access Service (RAS) A feature of Windows NT that allows remote users to dial in to and access the network. It is a versatile dial-up server service.

repeater Repeaters regenerate weak incoming signals. Because no packet information is necessary to perform this task, repeaters reside at the Physical layer.

RG-58 A type of coax commonly used interchangeably with Thinnet.

ring A descriptive term used for the logical path used in a Token Ring network.

router Connects different networks together.

Routing Information Protocol (RIP) Distance-vector routing protocol.

Serial Line Internet Protocol (SLIP) An older dial-up protocol that is being replaced by PPP.

Server Message Block (SMB) An Application layer protocol primarily used for file and print sharing.

Session layer The Session layer creates a virtual connection between two applications.

share-level security Security level in which each shared resource on the network has an associated password.

shielded twisted-pair (STP) Cable that consists of one or more twisted pairs of cables enclosed in a foil wrap and woven copper shielding.

Simple Mail Transfer Protocol (SMTP) A protocol used to transport electronic mail through internetworks.

Simple Network Management Protocol (SNMP) Network management protocol used with TCP/IP networks.

spooler Accepts print jobs and spools the jobs to disk until the printer is ready to accept data.

spread-spectrum radio transmission Type of radio transmission that changes frequencies regularly.

star Descriptive term for the physical layout of certain network types, such as 1OBaseT.

T connector Used in Thinnet-based LANs, this connects the network interface card to the cable.

Tl digital Line that provides point-to-point connections and transmits 1.544 Mbps in a total of 24 DS-0 channels. Also called a DS-1.

T3 digital Line that can transmit data at up to 45 Mbps. Comprised of 28 MUXed Tl circuits.

TCP/IP Typically used to refer to the entire family of Internet protocols.

TDR Time-Domain Reflectometer A device that sends sound waves to detect imperfections in a cable.

TechNet A CD-ROM-based repository of technical information available from Microsoft.

telnet A protocol that enables PCs and workstations to function as dumb terminals in sessions with host systems on internetworks.

terminator A special connector that includes a resistor.

Thicknet Also known as IOBase5, the Thicknet topology employs RG62 cable, which is much thicker and harder to work with than RG58.

Thinnet Also known as l0Base2, Thinnet is the most common bus topology. This topology employs RG58 cable, which has a 5-ohm impedance.

Token Ring A type of network that uses token passing.

topology The physical pathway in which computers are connected.

Transport layer The Transport layer ensures that packets are delivered in sequence and error free. The Transport layer breaks large messages from the Session layer into manageable packets to be sent out to the network.

uninterruptible power supply (UPS) Usually, a large battery that plugs into the wall and supplies about 15 minutes of power.

Index